CUMBRIA LIBRA

D1766317

3 8003 04332

Cumbria County Library
Carlisle Group H.Q.
The Lanes
CARLISLE
CA3 8NX

SERENGETI STORY

Serengeti Story

life and science in the
world's greatest wildlife region

ANTHONY R. E. SINCLAIR

OXFORD
UNIVERSITY PRESS

OXFORD
UNIVERSITY PRESS

Great Clarendon Street, Oxford, OX2 6DP,
United Kingdom

Oxford University Press is a department of the University of Oxford.
It furthers the University's objective of excellence in research, scholarship,
and education by publishing worldwide. Oxford is a registered trade mark of
Oxford University Press in the UK and in certain other countries

© Anthony R.E. Sinclair 2012

The moral rights of the author have been asserted

First Edition published in 2012

Impression: 1

All rights reserved. No part of this publication may be reproduced, stored in
a retrieval system, or transmitted, in any form or by any means, without the
prior permission in writing of Oxford University Press, or as expressly permitted
by law, by licence or under terms agreed with the appropriate reprographics
rights organization. Enquiries concerning reproduction outside the scope of the
above should be sent to the Rights Department, Oxford University Press, at the
address above

You must not circulate this work in any other form
and you must impose this same condition on any acquirer

British Library Cataloguing in Publication Data
Data available

Library of Congress Cataloging in Publication Data
Data available

ISBN 978–0–19–964552–7

Printed in Great Britain by
Clays Ltd, St Ives plc

Links to third party websites are provided by Oxford in good faith and
for information only. Oxford disclaims any responsibility for the materials
contained in any third party website referenced in this work.

CONTENTS

CONTENTS

LIST OF PLATES, MAPS, AND ILLUSTRATIONS

Plates

1. Wildebeest on the short grass plains in the wet season.
2. Wildebeest massing at Seronera before moving into the woodlands in June.
3. Ol Donyo Lengai, the remaining active volcano of the Crater Highlands that created the plains.
4. Short grass plains with kopjes. Crater Highlands in the background.
5. Moru Kopjes at the boundary of the plains and southern hills.
6. Nyaraswiga, the easternmost of the Central hills near Seronera in the savanna country.
7. The umbrella tree *Acacia tortilis* is the classic tree of the savanna.
8. Kopjes in the woodlands are special habitats for many animals.
9. Riverine forest along the Grumeti River in flood, the western corridor.
10. Banagi Hill, site of the first Warden's station in 1930 and the first scientific lab in the 1960s.
11. Entering data accompanied by the baby buffalo, 1968.
12. The buffalo followed Anna and I everywhere around our Banagi station.
13. A leopard found a warm spot next to Anna outside the tent.
14. We stalked the buffalo herd. A thousand heads came up to look at us.
15. Ruins of Fort Ikoma, built by the Germans in 1902.

Maps

Illustrations

ACKNOWLEDGEMENTS

Throughout my life there have been two people who have supported me. The first is my brother, Tim, who shared my life in the early years as we grew up in tropical Africa and with whom I have kept in close contact thereafter as we went our different ways, he and his wife Wilma to New Zealand and myself to Australia and Canada. The second is my wife, Anne, whom I shall call Anna. We were married at the age of 21 and she has effectively shared all my working life in Africa, and she appears frequently in this book.

Other people who feature largely in these stories include Mike Norton-Griffiths, a British colleague in our early years in Serengeti, an extrovert with a wonderful sense of humour and huge generosity. Another is John Fryxell, a tall wiry Canadian with great intellect. I first met him when he was an undergraduate in my ecology class of 1975 and we have been colleagues ever since. He conducted his doctoral work on the migrations of white-eared kob in Sudan, which features prominently in these stories. He has developed an international reputation as a mathematical ecologist, and he has taken over much of my research in Serengeti. Third is Simon Mduma, a quietly spoken and gentle Tanzanian who became my masters student in 1988 and continued after his doctorate first to run our research in Serengeti and then to direct all wildlife research in Tanzania as a senior member of the Tanzanian government. Ray Hilborn, at the University of Washington, has been one of my closest collaborators for more than thirty years. He is also a computer modeller of great skill, and although in his other life works on fisheries harvesting, where he has built an international reputation, he has worked with me on the issues of bushmeat harvesting by local peoples around Serengeti. There are of course many others who will make their appearance in these stories and we will meet them as we come to them. For reasons that will become obvious I have had to change the names of some of the participants.

Several people have helped me with the library work as background for the stories. Of the scientists in particular I would like to thank George Schaller and Hans Kruuk, who acted as mentors; Hugh Lamprey, who directed the Serengeti Research Institute; and Colin Pennycuick who guided my flying experience in the 1960s and 1970s. Mary Leakey helped me in the 1970s and 1980s and provided a refuge at a critical moment. Close colleagues Craig Packer, Andy Dobson, Sarah Durant, and Sarah Cleaveland worked with me in the 1990s and 2000s. Peter Arcese and his wife, Gwen, guided the research in the late 1980s and early 1990s and were heroes during the bandit attacks. Patricia Moehlman found Simon Mduma for me. All my students contributed to our understanding of the Serengeti, in particular John Fryxell, Holly Dublin, Greg Sharam, Ephraim Mwangomo, Ally Nkwabi, John Bukombe, and Grant Hopcraft. But above all is Simon Mduma, who has looked after and administered our Serengeti Biodiversity Program since 1996.

On the National Parks sides, Myles Turner taught me much about Serengeti when I first arrived. Kay Turner, Myles Turner's wife, has been a close friend over the decades; she provided important information on the history of Serengeti in the 1950s. Chief Park Warden (CPW) Sandy Field did much of the flying for me in the 1960s, as did Steve Stephenson in the early 1970s. CPW David Babu had to cope with the dark days of the border closure; we became close friends and he helped us in many ways with logistics of avgas fuel, vehicles, and food. Following him was Justin Hando about whom I write more in the stories, so I will simply say he kept us informed and consequently alive when the bandits were rampant. Of course there were many wardens and rangers who helped and protected us and I have not forgotten them. They worked under very difficult conditions with good humour. The Tanzania National Parks and Tanzania Wildlife Research Institute have generously given permission and help over the decades for our research. I thank the many Directors for their support.

Simon Mduma kindly read and corrected Chapters 8, 15, 17, and 18. Justin Hando provided valuable information, and read through Chapter 17. Erich Hinze generously translated the relevant portions of Oscar Baumann's account of his travels through Serengeti in 1891. Tim Sinclair helped with many of the stories. Gerald Rilling was invaluable in finding the early books on Serengeti. Ann Tiplady and Catherine Sease kindly provided the unpublished diaries and films of the Lieurance hunting expeditions of 1928–9. Sue van Rensburg and Norman Owen-Smith in South Africa provided the information on the roads through Hluhluwe and Kruger Parks for Chapter 20. I thank Rhodes House and the Bodleian Library at Oxford University for the use of their facilities, and the archivist Lucy McCann for help in finding material. Terry McCabe helped with the history of the Maasai. Kris Metzger has been working with me since her PhD in 1999, and in particular as a Research Associate since 2005; she has been invaluable in putting together the data and creating the maps. Peter Arcese and Frommann/Laif, Germany, kindly provided photos. Alistair Blachford and Andy LeBlanc at University of British Columbia (UBC) worked their magic to improve my own photographs. Also at UBC my colleagues Charles Krebs and Alice Kenney have helped so much with data analysis over the years; Judy Myers and Jamie Smith provided constant support.

Our work also depended on the generous help of friends in Arusha who provided a base for us. Bob and Kathy Gillis from the Canadian Wheat Scheme were our safe haven in the 1980s. Jo Driessen and Judith Jackson have provided a home for us since 1999. Our caretaker in Seronera, Juma Pili and his wife, Mama George, have worked for us since 1988 with unquestioning loyalty, even when JP, as we call him, was attacked by the bandits. Stephen Makacha first worked for me in 1967 and for many years thereafter; after two decades away he came back to work for us in 1998 until he retired in 2007; and now his son Saba has taken over. Joseph Masoy has been our most reliable and willing field assistant for nearly a decade. In Canada, I should mention Drs Stephen Kurdyak and John Mancini, who have kept me going all these years.

A scientist's work has to be paid for, of course. This is generally done by government grants to university researchers, contributions from conservation organizations, or occasionally private donations. For the great majority of my work in Serengeti, for some thirty-five years, the Canadian Government through their Natural Sciences and Engineering Research Council (NSERC) has funded much of this work, but other contributions have come from time to time through the Royal Society of London, British Ecological Society, Wildlife Conservation Society, New York, the National Geographic Society, the African Wildlife Foundation, Washington DC, the Killam Foundation, Canada, and the Frankfurt Zoological Society. The Friends of Conservation headed by Jorie Butler Kent stepped in at a critical time to keep us going. All of these organizations must take credit for the research we conducted but are in no way to blame for any of the events that I describe or errors in the telling. One of the main reasons we were able to keep the long-term research going was because of the policy of NSERC, Canada, which allows five-year renewable grants. This policy provided the security to develop the programmes that eventually told us what was happening in Serengeti and understand the causes. NSERC and my Canadian colleagues who reviewed the work deserve the credit; but so do the University of British Columbia, the Department of Zoology, and the Biodiversity Research Centre, who throughout my thirty-five years there provided support and understanding.

Lack of space and opportunity prevent me from mentioning all of the scientists and conservationists who have contributed to Serengeti and the greater East Africa, but this in no way belittles their important contribution to the enterprise over the decades—I value them all. In this context there were many people behind the scenes in Tanzania, Germany, the United Kingdom, and the USA who helped diplomatically when the northern road threatened the Serengeti ecosystem; in particular Guy Debonnet, Chief of Special Projects, World Heritage Centre UNESCO; Sir John Beddington, Chief Scientist UK; and my long-time colleague and friend John Krebs (Lord Krebs of Wytham).

The Frankfurt Zoological Society (FZS), headed by Markus Borner in Serengeti, has been the single most important funding body in keeping Serengeti going through the border closure and subsequent decades. Markus himself provided the leadership, and without him and FZS (and FZS Directors Dr Faust and Christof Schenck), I doubt Serengeti would have survived. Markus retired in 2012.

At Oxford University Press Latha Menon, Senior Editor, guided me and improved the manuscript immensely, while Emma Marchant has been invaluable with the logistics of communicating from half a world away. Thank you.

1

Serengeti: A Wonder of the Natural World

F AR in front of me were lines of wildebeest trekking slowly, each animal plodding steadily behind the other, the furthest larger than life, distorted in the mirage of heat. The lines stretched as far as the eye could see. They were moving across us from right to left as we headed east, line upon line disappearing into the haze, leaving the open plains for the woodlands. We had been driving for two hours now and for the whole time it had been the same picture of countless herds. Our track followed the edge of the Serengeti plains, a vast area of treeless grassland stretching 100 miles east of us to the foot of Ngorongoro and the Crater Highlands. The peak of the old extinct volcano, Lemagrut, rose to 8,000 feet, blue against the shimmering haze, forming the backdrop for ridge upon ridge of rolling plains. From time to time groups of zebra would lead a column of wildebeest, while the tan of a kongoni hartebeest stood out from the general grey and black of the herds. The kongoni were not migrating; they stayed behind, merely watching the herds pass.

The great migration was moving north; the end of the rains signalled the time to move. They stopped at the woodland's edge and slaked their thirst, for here was the first river, the Ngare Nanyuki. The water is alkaline, but the wildebeest can tolerate this. As we travelled we saw freshwater pools in which a few old male buffalo wallowed. Groups of Grants gazelles fed on the shrubs while their smaller cousins, the Thomson's gazelle, ran in countless numbers, stopping briefly to feed where wildebeest had been before. Grants, like the kongoni, were not moving, but Tommys were part of the great migration.

It was my second day in Serengeti. I had arrived late the day before, 1 July 1965, driving south from Nairobi with Sandy Field, the Chief Park Warden, across the Great Rift Valley in Kenya to Narok, above the western escarpment, and then south across the Loita Plains, through the Mara Reserve and finally into the Serengeti itself in Tanzania at the Klein's Camp gate. I was an undergraduate from Oxford and I was to study the migration of birds for a Royal Society-funded project under Professor Arthur Cain. Sandy deposited me at the tiny group of research houses at the foot of Banagi Hill and went on to his headquarters another 12 miles away at Seronera. There was no one at Banagi when I arrived, long after dark. A night watchman pointed to a rondavel, a small round hut, and told me to stay in there. I found a bottle of water and that was all I had that night. Next morning Onyango, the warden's driver, arrived at dawn and announced that I was to accompany him while he read the rain gauges—they all had to be read at the end of each month. We were already into July, but never mind. I left for the day. It was the first of three days of rain-gauge reading, and in that time we covered the whole of Serengeti, some 8,000 square miles, just Onyango and I.

We travelled east from Seronera along the northern edge of the plains until we reached a grove of majestic yellow-fever trees, green, cool, and peaceful, the Lerai forest at Barafu. From time to time we stopped to read rain gauges, pouring the monthly accumulation of rain into a measuring cylinder. The wet season had been late that year, and we had more than the usual amount of June rainfall, the reason why the migration was only now leaving the plains. From Barafu, a place of beautiful rocks (called kopjes in Serengeti), we headed south some 50 miles across grasslands so short they could pass as a golf course. The plains change from west to east. We had started in long grass up to our waists but as we drove east the grass species changed to shorter varieties, now only a foot high, and finally at Barafu they were very short, a mere 3 inches. As we drove across these lawns we passed tens of thousands of Tommys, feeding on the drying out grass, and they raced across our path as if to avoid capture. Flocks of plovers flew up disturbed by the running gazelles while all about us were small larks, pipits, and wheatears darting out of our way. Golden jackals and bat-eared foxes darted out of holes, while groups of hyenas lolled in the drying out waterholes. A solitary male lion basked in the sun oblivious to everything. We reached a lake, surrounded by a parkland of umbrella-shaped African acacia trees;[1] it was in a depression with a small scarp surrounding it. We looked down on Lake Lagarja; it was full, very alkaline, and covered in pink. This was the distant view of thousands of flamingoes. Thirty miles west, at the edge of the woodlands, we reached the Simiyu River where we turned north and went through a magical landscape of small rocky hills and kopjes, reaching up 100 feet and more, and covered in lush shrubbery and fig trees. These are the Moru kopjes, the home of hyrax, klipspringer, leopards, and baboons. It was after dark when we arrived home, having covered the whole of the northern plains.

Over the next three days I travelled with Onyango from rain gauge to rain gauge measuring the previous month's rainfall, across the whole Serengeti, some 8,000 square miles, over plains and savanna, forests and hills. The west was a patchwork of open grasslands between rivers lined with dense forest. To the south lies a hilly plateau edged with precipitous scarps, which forms a barrier along the western edge of the plains. The north has alternating rounded ridges with broad-leaved woodland, large evergreen trees in a parklike landscape, and steep valleys with forest.

Everywhere there were wildebeest—we were in the midst of what is now called 'the great migration', where herds in their millions move from the plains to the northern Serengeti at the end of the wet season. They stay in the north until the rains begin again in November, starting the great trek south. We met them that July spread across most of the woodlands, an area of some 5,000 square miles. But there were many other antelope species as well, some very rare, carnivores everywhere, and birds of countless varieties.

At the end of those three days I had seen the Serengeti as few had ever done. In the past I had seen something of East Africa, having been raised there, and had visited various game parks. But nothing had prepared me for this experience of wildlife in vast numbers, the extraordinary migrations, the sheer diversity of animals and vegetation, and the spectacular landscapes. I decided then that I would spend the rest of my life studying this ecosystem and why it was like the way it was. It was without a doubt in my mind the most extraordinary place on Earth.

* * *

The Serengeti is defined by the area across which the wildebeest migrate; it was first recognized as a place of global importance at an international meeting on national parks in Stockholm in 1972. The participants at that meeting agreed to set up the World Heritage Sites; these were areas that were deemed the most precious natural assets in the world to be protected against exploitation for posterity. Delegates drew up a preliminary list of the most important ones worldwide. The Serengeti came out top of the list. The name Serengeti is now a household word, the epitome of a wildlife spectacle in Pleistocene surroundings. Surprisingly, it has only recently come to be known thus. It was the lions that first attracted attention, in the 1920s—lions to be hunted by foreigners—and the wildebeest migration was completely unknown. The Serengeti plains were the place to go for the grandest black-maned lions in the world, and lots of them. Indeed, it was not until Bernard and Michael Grzimek from Germany flew their plane over the Serengeti in the late 1950s to document the great migration that the world first became aware of the phenomenon.

What makes Serengeti both unique and spectacular? What are the environmental features that allow a migration with so many animals? What determines the sizes of animal populations and the number of species that live in Serengeti? Indeed, why does it have so many species? These are some of the questions we will consider here about the biology of Serengeti. We will also look at its history, how it became a famous conservation area, and what might happen to it in the future. I first saw the Serengeti plains during the great floods of December 1961 but did not start my research until 1965. I continue it to this day. The Serengeti is significant because it supports one of the last remaining migrations of large mammals in a relatively unchanged state from the time of the hunter-gatherers, long before the agricultural development that surrounded Serengeti in the 1600s and before the impacts of the modern economic world. It is also a place of outstanding beauty and remarkable biodiversity—it supports more large-mammal species than any other place in the world. Despite its relatively undisturbed state, the ecology of the Serengeti has changed over the past century and these changes highlight its fragility and sensitivity to climate and human impacts.

Serengeti is a place where biologists can observe nature more easily than most. Its combination of open plains and savanna allows access to most of the area. The large animals are easy to observe. One can describe their ecology and behaviour mostly using only binoculars. Their populations can be counted accurately. Because of the many decades biologists have been studying the Serengeti they now understand the causes underlying the huge changes that have occurred in the ecosystem. In modern times human impacts on nature are becoming more severe and Serengeti has become a case study documenting these impacts. In essence the long-term studies have shown how political, economic, and social events have driven these ecological changes. Serengeti has become a valuable place to inform Science on how ecosystems work, and how they respond to pressures.

* * *

I have tried here to bring to life our endeavours to understand ecology in a tropical African savanna and how it has become relevant to human society and conservation. The story covers some fifty years. The many scientific papers and books that have been published about the Serengeti present the dry and sterilized account of our findings, reduced and distilled, for lack of space in journals, and for rapid transmission of results to other scientists who themselves do not have time to read more than the summary most of the time. However, science is conducted by people and is a reflection of their successes and difficulties in their endeavours. Scientific reports leave out the trials and tribulations, the failures as well as successes that underlie the science. It is with this in mind that I wanted to recount some of our experiences while recording data, making discoveries, and reaching important conclusions for society. This is also a book of stories.

During 1965 scientists conducted a census of elephant and buffalo and dis-covered that both populations were far higher than they had thought. These results were quite unexpected but matched the increase in wildebeest numbers that Murray Watson,[2] the biologist studying wildebeest, had already been documenting. Nobody could explain why any of these species should be increasing. So began the work to find out why the African buffalo popula-tion was increasing so fast.

2

The Great Migration

THE wildebeest migration across the Serengeti is one of the last intact migrations on Earth. There were once such migrations on most continents but humans have taken over the land and blocked the migration corridors, leaving the animals with nowhere to go. Many have perished as a result. Of the few migrations that remain, 'The Great Serengeti Migration' is the best known.[3] Serengeti is unique partly because of the special features of its geography that determine its environment, climate, water relations, and habitats. Together, these four create the migration.

Geography

Africa is splitting apart and in a few million years it will be two continents. The split is developing down a rift, the Great Rift Valley, from the Dead Sea in the Near East through Ethiopia to East Africa. In East Africa this rift splits into two (Map 1). The western arm, called the Albertine Rift, runs along the western borders of Uganda, Tanzania, and Mozambique. Within it lie the deep lakes Albert, Tanganyika, and Malawi. The eastern arm, the Gregory Rift, runs through the middle of Kenya and Tanzania. The edges of each rift are uplifted so that the land between the two forms a shallow basin. Lake Victoria is impounded in this basin, essentially a vast and shallow puddle of water only about 220 feet deep at its greatest depth. At 26,000 square miles it is huge, the largest lake in Africa, and the third largest in the world.

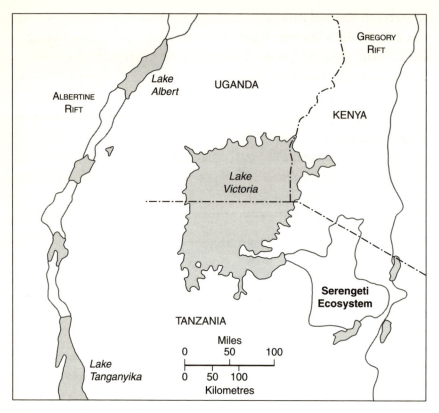

Map 1 General location of Serengeti, lakes, and the Albertine and Gregory rift valleys.

The Serengeti, as I shall call the whole system, is situated mostly in northern Tanzania, East Africa (Map 1); it is adjacent to that eastern spur of Lake Victoria called Speke Gulf, named after the explorer John Speke, who was the first European to see the lake and find the source of the Nile for European geographers. Of course he was directed there by local peoples who well knew of the lake's existence. It may not be an accident that maps of Africa as far back as 2,000 years ago already showed lakes in the centre of Africa giving rise to the Nile, and such rumours could have come from the Arab traders who were penetrating the interior then.[4]

The wildebeest migration covers an area of some 10,000 square miles and this includes many political administrations. The main ones in Tanzania are the Serengeti National Park (SNP) itself, and the Ngorongoro Conservation Area (NCA), which lies east of the park and includes half of the Serengeti plains. North of the NCA is the district of Loliondo through which wildebeest pass on their way north into Kenya. The Mara Reserve is the main Kenya administration. This holds the vital dry season grazing and water supplies. South and west of SNP are small game reserves, such as Maswa, Grumeti, and Ikorongo (see Map 2).

Most of the ecosystem consists of a flat or rolling landscape highly dissected with small seasonal streams that flow into a few major rivers. It is part of the high plateau of interior East Africa. This gentle aspect slopes from the edge of the Gregory Rift in the east to Lake Victoria in the west, so all the rivers flow west. The highest part of the plains is at 6,000 feet while Speke Gulf is at 4,000 feet.

There are three major rivers, the most important being the Mara, which originates in the montane forests of the Mau Highlands of Kenya. It has until recently flowed year-round (see Chapter 20), providing the main water supply for the great herds of migrating animals in the dry season. It flows through the Mara Reserve of Kenya and northern Serengeti, and eventually flows west through the huge Masirori swamp into Lake Victoria at Musoma. The two other rivers are the Grumeti, which originates in the highlands of northeastern Serengeti, and the Mbalageti, which originates on the Serengeti plains. Both are seasonal rivers with only pools remaining in the dry season. Two other rivers originate in southern Serengeti, the Simiyu and the Duma, but only the upper reaches lie within the Serengeti before they flow through agricultural land to Speke Gulf. All other rivers dry out except for a few springs that seep from the base of hills.

Steep rocky hills occur along the eastern boundary of SNP and between the Grumeti and Mbalageti rivers in the west, forming a backbone to the corridor between the rivers. The Nyaraboro Plateau with a high (1,000 feet) escarpment occurs in the south-west. Because of the generally higher elevation in the east the hills in Loliondo and the north-east of SNP reach 7,000 feet.

The ecosystem is effectively self-contained, cut off by natural boundaries on all sides. The eastern boundary is formed by the escarpment of the Gregory Rift and the base of the Crater Highlands. The south is bounded by

9

the edge of the Serengeti plains and in Maswa by the appearance of rocky kopjes. In the west the corridor, which is largely an alluvial plain formed by the rivers, is bounded on both its south and north by higher ground that is agricultural, and by Speke Gulf. The northern extension of Serengeti to the Kenya border is also bounded in the west by agriculture. Within Kenya the Mara Reserve is bounded by the Isuria escarpment, the Loita plains, and the Loita hills.

The environment

There are two special features that determine the Serengeti environment. First, the Crater Highlands in the south-east are sufficiently high (8,000 feet) that they impede the prevailing winds from the Indian Ocean, causing a rain shadow on their western side. The far eastern Serengeti plains, therefore, are semi-arid, receiving only 20 inches of rain per year. Indeed, there are sand dunes that are gradually moving across that region. The second important feature is Lake Victoria in the west. This lake is so large that it creates its own weather system; rainstorms develop over the lake and affect the west and north-west of the ecosystem, even during the dry season. So we have a wet northwestern region and a dry south-eastern region producing a marked gradient in rainfall. It is this gradient that drives the migration.

It is a highly seasonal environment with rainfall being the major influence. Rain is determined by the position of the sun. Serengeti is very near the equator, lying just 2° south. The sun passes over it on its way south in September and on its way north in March. Some six weeks after the sun passes, a band of heated air, the Inter-Tropical Convergence Zone, follows the sun and draws in wet air from the Indian Ocean. As a result, there are two wet seasons, a shorter one in November–December, and a longer one in March–June. Both seasons are variable, the shorter one often absent but sometimes the two merge to form one long wet season—as occurs when there is a strong El Nino event in the Pacific Ocean. There is a long dry season July–October. July is the driest month, and there are storms that become more frequent in subsequent months in the north-west of the system. In addition, Lake Victoria contributes rain from storms generated by heating over the lake.

Vegetation

We can think of Serengeti as having three major habitats that, by coincidence, lie along the rainfall gradient. First, there are the plains in the south-east. These are formed by a calcareous layer under the soil that derived from volcanoes long ago. This layer prevents trees growing and so the plains are open grassland. However, there is also a gradient of grassland types within the plains. The far eastern plains have very short grasslands, composed roughly of 40 per cent grasses and sedges, 40 per cent small flowering herbs, and the rest bare ground. All these plants are heavily grazed. They grow close to the ground as protection from being eaten. In the middle of the plains the grasses are longer, a foot high, mixed also with flowering shrubs—these are the intermediate plains. To the west and north lie the long grass plains. The Salai plains form the northern half of the eastern grasslands in the Loliondo district. They are very dry, and contain sand dunes, most of them covered with tussock grass, except some that are still moving.

The second habitat, the acacia savanna, starts abruptly at the edge of the plains as the effect of the volcanic soil disappears. The majority of the savanna is formed by many species of African acacias often as single species stands. Each has its own preferred position along a gradient of drainage. The gently rolling nature of the landscape results in well-drained ridge tops with sandy soils. The soil type changes progressively down the slope until poorly drained and even waterlogged silt soils are found at the bottom near rivers and drainage lines. The African acacia species lie along this gradient—umbrella trees (*Acacia tortilis*) prefer the tops, stink-bark acacia (*Acacia robusta*) prefers mid-slope, while yellow-fever trees (*Acacia xanthophloea*) like wet soils near rivers, and gall-acacias (*Acacia drepanalobium*, *Acacia seyal*) live in waterlogged swampy soils. These are just a few of the many tree species that occur here. The grass layer is composed of grass species similar to those on the long grass plains, but under trees there are other species together with many flowering herbs. The grass layer is much richer in species than on the plains. Along river banks and around kopjes there are shrubs which form thickets—favourite hiding places for predators.

The third major habitat is the broad-leaved woodland of the far north-west composed mainly of the trees *Terminalia* and *Combretum*. The soils are derived

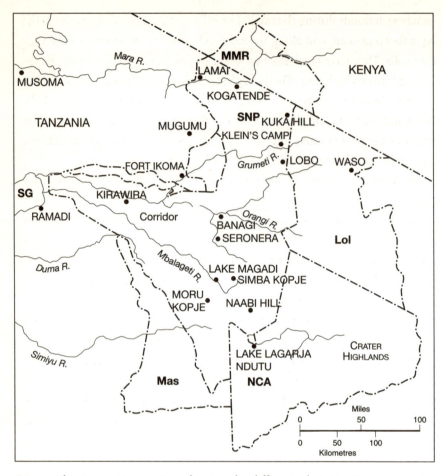

Map 2 The Serengeti ecosystem showing the different administrative areas, main rivers, and some locations mentioned in the text. SNP, Serengeti National Park; MMR, Maasai Mara Reserve; Lol, Loliondo Controlled Area; NCA, Ngorongoro Conservation Area; Mas, Maswa Game Reserve; SG, Speke Gulf of Lake Victoria.

from granite and poor in nutrients, the grasses are very tall, and there are many different species of shrubs.

There are also special habitats. The western end of the corridor was under Lake Victoria a few thousand years ago; it is now a flat floodplain—the Ndabaka floodplain—with alternating clay soils and sandy ridges (old beaches). It floods during the rains. The hills in the centre of the park, on the Nyaraboro plateau, and along the eastern boundary are stony or have very thin soils. They have *Combretum* on lower slopes, but a mixture of small acacias and shrubs higher up. On the higher hills of the north-east, such as Kuka Hill, the elevation results in montane forest, a relic of that on the Crater Highlands and the Loita Hills in Kenya. Most of this forest has been destroyed by fire over the past century and the hills are now covered by grassland; but small patches of forest remain in gullies where fire cannot reach.

The Mara River supports riverine forest, which is an extension of the montane forest downstream from the highlands. This is closed canopy forest maintained by a high water table from the river. It can be half a mile wide but more usually it is 50 yards or less in width. There is evidence it was much more extensive in past centuries. The Grumeti and Mbalageti rivers along the western corridor also support forest, but this forest is of a completely different type. It is a subset of the lowland Congo forest and has almost no overlap in tree species with the Mara montane forest.[5] So the Serengeti supports both major forest types in Africa, and only 35 miles apart.

The rocky outcrops or kopjes found in the eastern half of the ecosystem are granite intrusions, and are surrounded by a matrix of volcanic rocks. These small islands of rock support dense shrubs and broad-leaved trees such as marula (*Sclerocarya birrea*) and fig trees (*Ficus*). Vegetation is scarce on the kopjes of the eastern plains but lush on those at Moru, an area of large kopjes at the western edge of the plains in the south. Kopjes are small in area but are an important special habitat for animals.

The migrants

Every year during the great migration about 1.5 million wildebeest, together with 200,000 zebra and half a million Thomson's gazelle, move around the system—over 2 million animals in all. They all converge on the plains in the wet season because that is where the best food is. The grasses of the plains have the highest protein in the whole of Serengeti, but calcium and phosphorus are also high. The animals move around the plains following the rainstorms and the growth pattern of grasses.[6] The three migrant grazers, however, stick to themselves with only a small overlap in their distributions, taking advantage of the different heights of grass. As the plains turn green with the first rain, usually around December, the Thomson's gazelle arrive first, feeding on the short new growth. Then as the grass grows a little taller, say 6 inches, the wildebeest arrive and displace the gazelle who now move further east to the short grass plains. Eventually the zebra arrive and they confine themselves largely to the intermediate grass plains. This sequence of zebra in the western plains, wildebeest in the middle, and gazelle in the far east moves further east in lockstep as the wet season progresses and the plains become wet and waterlogged. They all move back west in reverse order when the plains dry out in May or June. However, in January–February there is quite often a dry period (between the two rains) and in that case the whole sequence moves south and south-west into Maswa Game Reserve—this Reserve is a vital retreat at this time of year. One other species also migrates. This is the eland, the largest of the antelopes. There are some fifteen thousand of them and they move onto the plains in the wet season. They feed on herbs, but also on grasses when these are green.

June sees the migration moving west and north. It is at this time of year that the herds graze the long grasslands. They move slowly—both because it takes time to graze long grass and because they are wary of predators—and so bunch up and form the dense masses that have become famous from photographs. Wildebeest dictate the movements of the other species. They eat down the grass and provide a niche for the gazelle that follow behind.[7] Zebra like to stay with wildebeest because they are safer there, but they must stay in front because they need a greater bulk of food than the wildebeest. So the great herds are seen with a front fringe of zebra, a mass of wildebeest, and then a dispersed scattering of gazelle behind.

Once the herds reach the woodlands this pattern breaks up and smaller groups of wildebeest and zebra make their way west and north during July to August. Thomson's gazelle stay behind in the central woodlands and by the later dry season there is little overlap with the other two species. By the end of the dry season (September, October) the wildebeest are in two major groups, one in the corridor, the other in the north-west of Serengeti and in the Mara Reserve of Kenya. Eland move north to the *Terminalia* woodlands where they feed on the more abundant shrubs.

14

The beginning of the rains in November brings the migrants south and east again towards the edge of the plains. However, the rains usually begin with scattered thunderstorms and this causes the herds to spread over most of the woodlands searching for the local patches of green food. They are at their most dispersed at this time. They only congregate again when the rains become more consistent and the herds move onto the plains.

Wildebeest have evolved several remarkable adaptations to their migration habits over millions of years. Such large numbers of animals are naturally the target of predators, especially the two dominant species, lion and hyena. When wildebeest give birth, predators converge on the calving grounds—the babies are easy to see and easy prey on the short grass, open plains. However, wildebeest synchronize their births, producing most of them—about half a million—within three to four weeks. Predators can only eat so much in a day, no matter how easy the calves are to catch; once the predators become satiated the remaining calves escape capture. But how can wildebeest synchronize their births? To do this they have to synchronize their conceptions nine months earlier in May and June. How they do this was one of the surprises that came out of the research I describe in Chapter 9.[8]

A newborn wildebeest calf has several adaptations to escape predation. It can run within fifteen minutes of birth, and run as fast as its mother within twenty-four hours. So if it escapes capture at birth, it can often outrun the predators when they start hunting next day. Secondly, mothers can time their births so that most of them happen between 9 a.m. and 11 a.m., which is immediately after hyenas have stopped hunting for the day and allows the maximum time for babies to gain strength before the next hunting period that night.[9]

The herds are always on the move and the babies must follow their mothers from birth. Both mother and baby learn rapidly who is who. However, stampeding herds can result in babies losing their mothers. In those thousands of animals there would be small chance of their meeting up again—and females never suckle calves that are not their own. But they do have a way of meeting: calves that are lost do not move along with the lines of animals; instead they stop and eventually they are left behind in the open. This makes them obvious to any predator around, but it also makes them obvious to the mother. The mother also does not move along with the herd. Instead she resolutely moves in the opposite direction until she comes to the back of the line—and there she can find her calf because there are now only a few to choose from.[10]

The residents

The distinction between those we call migrants and those we call residents is more one of scale than absolute differences. All species move with the seasons, but whereas the migrants move some hundreds of miles, the residents usually move only a few miles.

Topi and kongoni are both close relatives of wildebeest, being in the family called Alcelaphinae, and they all eat grass alone. Both species are sedentary, topi living on wet grasslands in large groups of several thousand in the corridor, while kongoni are in the eastern woodlands and long grass plains where conditions are much drier. Both species overlap in the centre and northern woodlands, living together in small groups.

The commonest antelope is impala, the quintessential animal of Africa. Paradoxically, it is the only one of its kind; there are no close relatives. It lives in herds of 10 to 200, all females and youngsters, and one male. All other males are forced out to live in large bachelor groups.[11] They live wherever there is savanna, feeding on shrubs, herbs, and green grass, and never venture onto the plains. There are several species of antelope that live in more restricted habitats. Water-loving antelope include waterbuck (larger than wildebeest), and the smaller common reedbuck, which is abundant in all tall grassland but rarely seen except along rivers. There is also the rare mountain reedbuck, found only on the top of the highest hills. Bush or forest antelope include bushbuck wherever there is thicket along rivers, and greater kudu in the kopjes of Maswa. Lesser kudu are found in the montane forests of the Loita Hills, which are not part of our system, but it is possible they are in the forests of the Serengeti highlands in Loliondo, yet to be discovered.

On the Salai plains we find beisa oryx. They are rare and their status is currently unknown. Related to these are roan antelope. They prefer, strangely, the low-nutrient granitic savanna typified by broad-leaved woodland. They were once abundant over the whole of northern Serengeti and Mara, some

associated with the central hills such as Banagi, but they have since disappeared. One small group is being nurtured in the Grumeti Reserve and
another in the Maswa Reserve. Another species that prefers the broad-leaved
woodland is oribi, a small antelope that lives singly or in small groups.[12]
Other small antelopes include grey duiker in similar broad-leaved habitat
and steinbuck that prefers dry acacia savanna.

Rhino were once common over the whole Serengeti, including Olduvai
Gorge. But in 1977 rampant poaching effectively exterminated rhino and only
a few remain in the south and in the Mara Reserve. Careful guarding of these
few have kept the population alive since 1990 but they have not increased
much, and efforts are now focusing on artificially adding to their number in
the hope that they will breed faster. African buffalo were also abundant,
reaching 70,000 in the mid-1970s, but the same poaching reduced their numbers to 20,000. They are slowly increasing again. They form large herds, up
to 1,000 strong. They live in the savanna, feeding on long grass, and need
daily access to water. The north-west is their optimum habitat although they
are now rare in that area.[13] Giraffe are also ubiquitous in the savanna. Numbers are not well known although we think there are about eight thousand.
They may be declining due to poaching.

Carnivores

Serengeti was first famous for its lions; foreign hunting expeditions camped
at the edge of the plains on the Seronera River because there were so many
lions in the vicinity. It was the extermination of lions that led to the first small
part of Serengeti becoming a protected area in 1930. The Seronera River is
still the best place to observe lions, although they occur over the whole ecosystem.[14] Leopards also live throughout the savanna and can also be found
around the kopjes in the long grassland wherever there is dense vegetation.
They are always secretive and difficult to see. Cheetahs are seen most
frequently on the plains but they occur in the savanna also. There are also
several small cats—serval, caracal, wildcat—but they are not often seen.

Hyenas are the most abundant carnivores in the system, some seven thousand of them.[15] They live in groups on the plains and hunt large ungulates

together. They follow the migrating herds around the plains and some way into the woodlands, perhaps 30 miles. Once the herds get beyond this, hyenas that live on the plains do not follow. There are hyenas living throughout the savanna, hills, kopjes, and even forest; these are usually solitary and scavenge or feed on small animals. Hyenas that live in packs are also found on the larger grasslands within the savanna, such as those at Musabi and Togoro.

Several species of the dog family are also found in the Serengeti. The best known are wild dogs, which were once found throughout the savanna and plains. Strangely, European hunters and wardens disliked them and shot as many as they could—despite the area being protected. Then starting in the late 1960s wild dog numbers within the ecosystem began a decline (linked to the increase in larger predators and the spread of disease from domestic dogs, which I will describe later) that ended with their disappearance in 1992. A few packs have since been discovered on the very eastern edge of the ecosystem—they had probably always been there undiscovered. They have not reinvaded the central savanna and are regarded as highly endangered. There are four other dog types. The black-backed jackal is common in the savanna, the golden jackal is found mainly on the short grass plains and around Ndutu, and the side-striped jackal is solitary, seen only occasionally in both long grass plains and savanna. The bat-eared fox specializes in eating dung beetles—their large ears allow them to hear beetle larvae underground. They live in holes throughout long grass plains and savanna.

The Serengeti also has many species of small carnivores belonging to the weasels (family Mustelidae—striped weasel, zorilla, honey badger), the mongooses (Family Herpestidae—six species), and genets (Family Viverridae—civet, common and spotted genet), most of them feeding on rodents and insects.

Afrotheria

Recent taxonomy using comparisons of genomes has shown there is an ancient group of mammals that evolved when Africa was cut off as a great island, some 100 to 50 million years ago, at the very beginning of the age of mammals. They are called Afrotheria, and they include the elephants, hyraxes, aardvark, and elephant shrews: a truly African group.

Bush elephants—those that occur in the Congo forests are now considered a separate species of forest elephant—occur in savanna and long grass plains and number about three thousand. They could be considered as migrants. There are two and possibly three populations, a northern group based on the Mara River and its forests in the dry season—including both the Kenya and Serengeti portions—and a southern group based on the Duma and Simiyu rivers and rivers in Maswa. In recent years a third group has appeared in the western end of the corridor based on the Grumeti and Mbalageti rivers. The northern group moves south some 100 miles towards Seronera in the rains. The southern group moves east about 50 miles to Moru, and even across the plains to Lake Lagarja when it is wet. The western group appears to remain more sedentary in the corridor.

Hyraxes are rabbit-sized animals that live in kopjes and forest trees. Two species—rock hyrax that feeds on the ground and bush hyrax that feeds in trees—live together in the crevasses of kopjes throughout the savanna and a few kopjes on the edge of the plains where there is enough shrubbery. The third species, tree hyrax, lives only on the Mara River in holes of large riverine trees.

Perhaps the most unlikely relative of the elephant is the aardvark. With its strange tubular teeth, long snout and tongue, and long digging claws it is supremely adapted to eating ants and termites. It occurs everywhere there are termite mounds. It digs into these from the side to access the nests using its long tongue to lick up the insects. It is nocturnal, solitary, and rarely seen. But its influence on the ecosystem is large, for many species use the holes as their houses.

Other animals

Among other animals, biodiversity is very high in some groups. There are over six hundred species of birds because of the high diversity of habitats, and many are influenced by the wildebeest migration. For tourists, birds are one of the more obvious features of the Serengeti. In contrast, rodents are not obvious but they are also diverse; some thirty species occur, and many of the small carnivores and birds of prey depend on them.

Insects are not yet well described, but we have some 180 species of butter-fly, 100 species of dung beetles, and 70 species of grasshoppers. The insect fauna supports the majority of the bird species. In contrast, both the reptile and amphibian fauna are not diverse. There are only a dozen species of liz-ards, some twenty of snakes, and thirty of amphibians.

This is just a rough outline of the immense richness of the Serengeti ecosystem. The variety of species is very large, encouraged by the different habitats and the impact of the great migration itself. This is the Serengeti as it now stands. It has arrived in this form through a long series of changes caused by events over the past 200 years and earlier. How these changes have come about and what their impacts have been on the biology of Serengeti unfolds in the story I now recount of the scientific research and conservation over the past fifty years. I start with the first question: what was causing the increase in the buffalo and wildebeest numbers in the early 1960s?

3

African Buffalo

SINCE the early days of the twentieth century scientists have been asking what causes animal populations to change in number and what prevents their extinction. These were the most hotly debated questions in population biology for some fifty years.[16] One theory stated that when environmental conditions were good survival was also good and so numbers increased. The reverse held when conditions were bad. A population did not go extinct because the environment changed in time to allow an increase again. Such an idea was contested on the grounds that changes in the environment depended on random events of weather alone and eventually extinction would occur. Other theories therefore proposed that changes in food supply, predation, or disease were involved in addition to changes in weather. By the mid-1960s there were plenty of theories but a dearth of evidence from real cases in nature, which was required to resolve the arguments. But the causes of population change are difficult to pin down for they require a detailed knowledge of both the demography of the animals—their births and deaths—and the factors that influence these, their food and what kills them. At that time most of the animal species under study were small—pest insect populations in Australia and bird populations in Britain. When such animals die they simply disappear. There is little trace left to show the cause of death. So the arguments were difficult to sort out.[17]

In 1965 we found that African buffalo numbers were increasing at a high rate, as were elephant and wildebeest numbers. Buffalo are very large animals. They are easy to see and so easy to count and record the births; when they die their carcases remain for a long time and so can be examined for causes of death. They are large enough for us to observe directly what they were feeding on. Once we knew their food we could measure their food supply. Here was an opportunity to find evidence for what caused changes in populations. In particular, we wanted to know why this population was increasing; the answer was directly relevant to their conservation.

* * *

The first step was to observe what buffalo ate. It proved to be not quite as simple as it seemed. In the 1960s there was very little known about the ecology of buffalo, including their food, and certainly not in Serengeti. Sometimes it is possible to get up close to animals and watch what they are feeding on. This we could do for some antelopes such as Thomson's gazelle or topi. However, trying to do this with wild buffalo proved to be a problem in Serengeti. Other scientists[18] had been able to observe buffalo herds close up where they had become very tame in the Queen Elizabeth Park, Uganda, but in the Serengeti the animals were very timid, unused to disturbances by humans, and the herds ran away when we were half a mile away. All we saw was a cloud of dust.

We had become aware that grazing animals, such as buffalo or wildebeest, distinguished between different components of a grass plant. They were highly selective, some preferring to eat leaves at the base, or leaves up the stem or the stem itself. To look at such fine choices we had to get very close indeed, just a few feet away. It therefore became obvious that we had to have some tame buffalo; then we could follow and watch what they chose to eat. With that knowledge we could measure what was available as food for the herds in the park.

* * *

We had the precedent of Oliver, of course. Oliver was a male wildebeest. He had been raised by Murray Watson[19] in 1963 and his job was to teach humans what wildebeest liked to eat and drink. He did this admirably while he was small and all he was interested in was food. However, things changed when hormones kicked in and he became interested in other wildebeest. As he grew up he became more and more of a problem. Wildebeest have a number of interesting adaptations, which we have seen in the previous chapter, but none of them involve living in harmony with humans. He thought he was a human, or rather he thought all humans were wildebeest. When he became mature he challenged every human he met to a duel of horns, and since humans did not have horns, he won every duel hands down. Wildebeest fight by charging at each other and then, at the last second falling to their front knees, colliding horns, wrestling and pushing, until one gives up and runs away. Humans gave up right away and ran, closely followed by Oliver horning the unfortunate surrogate in the backside. Then he would stand, presenting horns, as dominant males do when they have vanquished an inferior intruder to their territory. He became obstreperous, even cocky, having won every bout. He would stand in the middle of the courtyard of Banagi, the small research station where all research in Serengeti began, the centre of his territory, looking around for challengers. Those driving in had to run the gauntlet to reach their houses before being hoisted, which led to much protest, but it was to no avail.

If there were no researchers to challenge and all else failed there was always the cook. Banagi house had a kitchen detached from the main building, situated some ten yards behind and joined by an open walkway to the back of the dining room. The cook had to carry the meal along this corridor, laden with trays, plates, and bottles. It was a favourite entertainment of Oliver's to lie in wait at lunchtime, hiding behind the building and then darting out to upend the hapless cook, who dropped the meal and fled back to the kitchen. By 1966 Oliver had become a nuisance, and without his sponsor, Murray Watson, who had left in February 1966, his welcome was over.

* * *

It was clear that our buffalo were not to get out of hand. But first we had to catch one. At first sight capturing a baby buffalo seems easy. Newborn animals can barely run. In the 1960s when I drove towards a herd of wild buffalo they would turn and run, maybe not very fast, say 30 mph, but the babies ran even slower and they were left at the back, trailing behind the herd. So one merely had to drive up to a baby, stop, pick it up, and push it into the back of a pickup where helpers would tie up the legs and off one goes home. Well that was the theory. It turned out to be somewhat different in practice. True the babies were left behind and we were able to drive up to them, but they kept running, and someone had to jump off the pickup and run with them to bulldog them, that is throw them to the ground, tie their legs, and hoist them into the vehicle. Buffalo babies are already 50 lb in weight and put up quite a struggle, which inevitably meant the bulldogger fell to the ground while trying to tie up the baby. Most scientists are not trained as cowboys so the whole procedure was usually a shambles. All the while the baby was bellowing at the top of its lungs, which was a signal to the mother to turn around and come to the rescue. Writhing around on the ground trying to disentangle oneself from buffaloes and ropes is no place to be when an angry female buffalo of 800 lb is charging flat out with horns well adapted to tossing predators into the air. We soon found that there was very little time to disable the calf before we had to make our escape and it took many attempts—and several narrow escapes as the mother hooked the back of the vehicle—before we succeeded.

In the end we captured two newborn animals a few months apart. The first one, named Bogo, was a female. She put up a fierce fight as we drove her home to Banagi, bellowing all the way, and mother following close on our heels—indeed I wondered for a while whether we would have her still with us when we got back to base. Eventually she gave up and returned to her herd. At Banagi we had a stall prepared where we released her. She promptly charged my legs, knocking them from under me and leaving me prostrate while continuing to bellow for her mother.

I had to feed her. I prepared a large bottle of milk from dried milk powder used to feed cattle calves and with the help of an assistant held her between my legs and put the teat in her mouth. Holding her in this fashion resembled how they suckle in the wild, which for buffalo was to put their head between the back legs of the mother and butt the nipples. As soon as she tasted the milk she calmed down, took the whole bottle and then another until she was replete, and from that time on she never attacked me again. She settled down to sleep for the night and next day she followed me around as her mother and to the best of her ability never let me out of her sight. A few months later we succeeded in capturing Rudolph, and the same remarkable transformation of behaviour took place with the first feeding.

* * *

I came, therefore, to have a family of babies that followed me around wherever I went in our small homestead of Banagi. This little station involved a main house built of mud in the 1930s for the first warden, a small house at the back, which was mine initially, and a rondavel for visitors, a generator shed some fifty yards away, and a laboratory built by Professor Grzimek in memory of his son Michael who was killed in an air crash in 1959. There were also four or five houses for field assistants, mechanics, and caretakers. It was 12 miles from the nearest habitation in the middle of Serengeti.

The baby buffalo roamed at will around this encampment perfectly habituated to humans, feeding peacefully. In the heat of the day they knew that I went inside our house, which had rooms with cool smooth concrete floors and deep cool shade. When I was there the buffaloes would try to follow but I would keep them outside and they made the best of whatever shade they could find by the house, peacefully sitting and ruminating. From time to time they would wander into the house if I was not there because that's what one does at that time of day. The first time I came across them peacefully settled in the middle of my sitting room I yelled at them, waving my arms to frighten them out of the house and to teach them not to come in. This was an unwise thing to do for they immediately jumped to their feet, hooves sliding in all directions on the slippery concrete floor; they were literally spinning their wheels, and in their fright they did what all animals do. They defecated and urinated everywhere. Eventually they made their way out and I was left to clean up the mess. They never did learn that where I sometimes went they should not go, but at least I learnt that when I found them inside, which was quite frequently, I should coax them outside again very gently.

* * *

In the evenings my wife Anna and I were in the habit of taking a walk for a mile or so around the station to look at crocodiles or hippos in the river and birds in the nearby bush. Naturally the buffaloes came too. It soon became apparent, however, that there was a particular order in which we had to walk and no other was tolerated. If Anna and I fondly thought we could walk together hand in hand, we were soon disillusioned. In buffalo herds babies follow their mothers and that is the way it is. When I went for a walk my babies followed me, the older one, Bogo, directly behind me and the younger one, Rudolph, behind Bogo. If Anna tried to join me she was brusquely horned aside; there was no tolerating her in front, and indeed she had to walk number four in file astern. Provided we obeyed the rules the buffaloes were very happy to go for a walk and they often ran ahead, gamboling as cattle calves do with their tails up and curled forward, kicking their heels in the air; only then could Anna and I walk together. But once they came back to our path it was back to the set order of things. If, however, Anna took off for a walk on her own they were happy to follow her as the surrogate mother.

* * *

As part of my work on the wild buffalo I collected the skulls of animals found around the Serengeti to record the age at which they had died. Some of these skulls of adult males were impressively large; some were wider than anything in the hunter's record books. We lined the driveway up to our house and the edge of the Banagi courtyard parking lot with these skulls and after a while they became something of a tourist attraction. This meant that we were sometimes overwhelmed with groups of tourists wandering around our house and photographing the skulls, and sometimes they even came into our house. They became something of a nuisance.

On one occasion tourists were swarming around the front of our house photographing the skulls. Rudolph was around the back of the house where there was a very high stack of dead wood, some 8 feet high, used as firewood to heat the hot water for the house. Rudolph's long rope trailing behind him became snagged in one of these logs and Rudolph gave the rope a mighty heave to free it, and brought the whole stack crashing down. Startled, he ran in panic, head forward, mouth open, tongue curled out, bellowing as loud as he could. He appeared around the side of the house making straight for the tourists, the log dancing behind him. The tourists' first view was of a buffalo

26

charging straight at them making a banshee racket, followed very shortly by their legs being knocked from under them by rope and log. In an instant there were bodies lying all over the place, while the remaining terrified onlookers ran in all directions into the bush. They fled, never to return.

* * *

The population work required camping out at various locations in the park. Into our Land Rover we bundled camp gear, food, and our camp assistant Makanga, for a week or so of bush camp. This was our first Land Rover. It had four wheels—sometimes, a canvas top which was ripped, doors that would not close, and an engine that was temperamental, preferring to stop without encouragement at inconvenient places. It caused us no end of trouble.

While travelling south along the north-western boundary of the park, a very rough and remote track, the radiator sprung a severe leak, far too big to fix with the normal tricks of pouring in cayenne pepper or egg albumen. There was nothing for it but to keep filling the radiator—or rather keep pouring water in as fast as it leaked out, which was fast. So it was that we travelled with the hood off, I sitting on the edge of the engine with a jerry can pouring water into the radiator while Anna drove. This took a lot of water and we had to drive from waterhole to waterhole, filling up our cans at each. It was touch and go that we reached the next source of water before we ran out, roughly 5 miles. We took two days to cover the 100 miles or so home, hot, dirty, and tired.

Some six months later we were camped in the southern end of the park, at Moru, where there are magnificent kopjes covered in dense and lush vegetation. Fig trees provided shade for our camp, fruit for birds, and wonderful resting places for leopards. It was a little patch of heaven. Makanga was there to look after the camp, long suffering as ever, for he really did not like to be out in the bush.

One day Anna and I had to return to Banagi to pick up supplies and fuel. It took us longer than we thought so it was eleven at night before we got back to camp. It was strangely deserted, no sign of Makanga even in his tent. We called out for him. There was an answer—from high in one of the fig trees. He answered, as he clambered down, that he had been hiding from the lions.

We had not seen any around but he had decided to do so just in case. Enquiring why he did not just stay in his tent, he replied that the tent walls were no thicker than his shirt and were not likely to fend off lions. This is an argument that rangers express to this day, many preferring to sleep outside a tent and by a fire, where they say they can see the lion coming. I tried to reassure him that lions don't come into tents, but he reminded me of an incident involving Jon Wyman, who had been studying jackals when I first arrived in 1965, and whose tent fell down as lions went past (see Chapter 7).

Makanga may well have had a point. A few months later, in June 1968, Anna and I were camped at the eastern end of Lake Magadi, near Moru, in a beautiful stand of yellow-fever trees—now long gone. We were sleeping on foam mattresses, and Anna, in her sleep, had rolled to the side of the tent. I was awoken by her urgent whisper that there was something outside the tent. Befuddled, I mumbled that there were lots of things outside the tent. No, she insisted, it was beside her on the other side of the canvas, she could feel it growling. I listened and sure enough I could hear a distinct purring, like a domestic cat only much deeper. 'How big is it?', I asked, to which she replied that it was much bigger than a cat, and it was leaning against her. It was clear that whatever it was had found a nice warm place next to the tent wall and was now comfortably resting against it. It was also clear that it was perhaps wise not to frighten it in case it attacked the tent wall. You had better not move, I suggested, it thinks you are a nice warm pillow. So she lay, motionless—and sleepless—until it decided to depart an hour or so later. Meanwhile I had fallen asleep much to her annoyance.

Next day we looked at the spot. The spoor of a leopard were clear to see.

* * *

As part of our studies on the population changes I needed to know how many calves were added each year. In the early stages of the work I tried to do this from the ground, assigning sex and age classes to each animal in the herd. However, to do this required getting up close and, as I have mentioned, the Serengeti herds were very timid so that they ran away when they saw a vehicle. Undaunted I decided that the best way to get close was to stalk them on foot, creeping from downwind through long grass to a convenient termite mound. From there I could count off each animal. The problem was that I needed an assistant to write down what I was counting since there were as many as a thousand animals in a herd and I could not take my binoculars off them to write.

It was February 1967, and a young lad, Patrick Duncan, just out of high school had come out for some experience and had been assigned to me as a field assistant. Patrick was an enthusiastic, indeed exuberant, young biologist, willing to learn anything. We were in northern Serengeti near Wogakuria where there were many large herds. The countryside was of rolling hills and valleys, open long grassland with patches of dense thicket, largely of *Croton* bushes. They were the favourite haunts of rhino and elephant. I told him what we were about to do. He was nervous; stalking buffalo was the sport of big game hunters with huge rifles, not that of unarmed boffins with binoculars facing a thousand of them. We stalked the herd, getting to within about seventy yards, and from a convenient hillock I read off the animals while Patrick sat behind the mound writing it all down.

I pointed out to Patrick that there was really nothing to be frightened of— they were grazing just like cows. If I stood up and shouted they would all run away, I announced. Patrick considered this to be foolhardy in the extreme; he would have rather that we crept away and then breathed a sigh of relief when this ordeal was over. Tempting fate was asking for trouble—what if they all charged together with us out in the open and nowhere to go? I thought he should learn a thing or two about buffalo, so undeterred I stood up on the termite mound, clapped, shouted, and waved. A thousand heads came up in total astonishment, paused for a few seconds as they tried to figure out what this strange apparition was, and then turned and vanished in a huge cloud of dust. See, I said, turning to Patrick, but I found I was talking to no one. Patrick had not waited to see anything; he had run for cover, which in this case was one of the *Croton* thickets some fifty yards away. He was already there as I shouted not to go in there, it was dangerous. But too late he disappeared into the thicket, only to reappear a few seconds later, running back to me even faster. Behind him was a rhino charging and snorting. Now we *were* in trouble, and we both had to find protection. There were a few large trees scattered about and we raced for these, each of us swinging up into one as the rhino arrived. And there we stayed, perched one in each tree like a pair of baboons, calling to each other, with the rhino huffing and snorting below us. He could smell us and hear us, but he could not see us, which worried him a lot. We waited until he eventually gave up, lost interest, and wandered off. Patrick learned fast and became a valuable assistant. He later went on to study the biology of topi in western Serengeti.[20]

* * *

George Schaller was the lion biologist who had arrived in early 1966 having already completed impressive studies on mountain gorillas and tigers.[21] His was the first of what was to become to this day an almost unbroken series of studies of the lions of the Serengeti plains. Late one day in December 1966 George was near Moru kopjes on the upper reaches of the Mbalageti River. There in a swamp was a pride of lions, which for some reason he needed to get close to, so he drove into the swamp. It was a bottomless swamp and in no time he was up to his axles in mud with no hope of getting out.

It was some 30 miles back to his house at Seronera where his wife, Kay, was waiting for him. It was too late in the day to walk home and it was not wise to be walking across the plains at night, unarmed as we all were. So he stayed the night in his vehicle surrounded by lions. Next day he somehow got out of his Land Rover despite the lions and set out walking home. Meanwhile, his wife had reported him missing to the Park Wardens after he had failed to return the night before. Myles Turner took off in the Park's Cessna 180 aircraft to search for him, and to his concern saw the vehicle in the swamp, empty but surrounded by very content and somnolent lions. He returned to base and he and the Chief Park Warden, Sandy Field, then conferred on how to break the tragic news to Kay. Myles then went to the Schallers's house to tell Kay, knocked on the door, and found to his amazement that it was opened by George—who had just got back from his long walk home.

This story did the rounds of the community with much hilarity so I was well aware of it when in January 1967 I was also at Moru, which was one of my special areas for buffalo herds. I was not aware of the swamp in which George had become stuck. Late one evening I saw a herd of buffalo across the same river and thinking the crossing looked easy drove in—and promptly bogged irretrievably to the belly of the Land Rover. It was indeed the same swamp. It was 6 p.m. and far too late to walk, so I had to stay the night, feeling foolish since I had not learnt a thing from George's escapade.

I had with me the tape recorder which I used to count off the animals in a herd, and a tape that I had borrowed from Hans Kruuk.[22] I entertained myself by listening to the tape, on which Hans had recorded his report on his studies of hyena. Hans was the first to record the biology of the spotted hyena, dispelling the many myths that then abounded. It became a classic study that is still outstanding today. From this tape I learnt that hyenas hunted mainly at night, starting at dusk and continuing to 8 a.m. when the sun became too hot to hunt. I also learnt that hyena hunt in packs for zebra but in smaller groups for wildebeest.

It was an unpleasant night. I alternated between staying in the vehicle in the swamp where I was bitten relentlessly by mosquitoes until it was unbearable; or wading to dry land, making a fire and staying in the smoke which deterred the mosquitoes. Amongst the stories of the hyenas I had heard that they were not much put off by fires and were quite happy to pick off sleeping people—whether this was true I was not sure but I was not going to put it to the test. Tiredness overtook me so that I was forced back to the vehicle where at least I felt safe.

The important decision was when to set out to walk. If I waited until 8 a.m., when hyenas stopped hunting, there was not enough daylight left to walk to Seronera. If I left earlier, at daybreak near 6 a.m., I ran the risk of meeting hyenas on the open plains with no cover. In the end I opted for the latter, setting out at 6.30 when at least it was cool. I had an hour and a half to endure until relative safety.

All went well for the first half hour; I was enjoying the sunrise and the birds calling. Then, ominously, I saw a hyena in the distance. Never mind, I reassured myself, it was a loner. Well not quite. Another appeared, and then another seemingly from nowhere, and presently there were five of them. They came towards me, I was not sure what to do, I was on the open plain with no protection for 10 miles or more. I knew I should not run; this would cause them to chase me. All I had was a hunting knife, which was useless in this circumstance. So I walked on and in due course they surrounded me, getting ever closer. I thought morbidly that they were treating me like a zebra, not a wildebeest. Now I knew what it felt like.

Whichever hyena was behind me rushed at me so that I was forced to turn round and round while I shouted at them to keep them away. It was not working. They were almost on me now, and their tails were up, a sign of the hunt. I felt the hair on the back of my neck tingling, my chest and stomach began to tighten; primeval fear was overtaking me. I could not allow them to grab me; it would then have been all over. They were now only 3 yards from me. Time seemed to be going in slow motion. I had to do something.

Rarely on the plains are there stones. But I noticed in my desperation that here were some small stones. I stooped to pick some up and as I did so they rushed at me. I leapt up and threw the stones, some hitting most missing. The ones that hit did nothing to deter them. The ones that missed bounced behind them—and strangely this seemed to upset them; they turned to see what it was and retreated a little. It did not take long for me to repeat this deliberately, and miraculously it worked. I replenished my stock of stones and lobbed them behind.

I had also spotted a small rock, some 3 feet high and about as much across, about a quarter of a mile away, the remnants of a kopje. I walked sideways towards this, slowly, very slowly, all the while lobbing stones. Presently, I reached the rock and climbed up. At least I was now much higher than them, and although they could easily reach me, my height seemed to intimidate them. They stood around and then lay down. I was not going anywhere. I was going to outlast them if it meant staying all day, now that I felt a bit safer. It was a stand-off and so we all waited for some two hours. The sun got hotter and hotter—it was the hot season and it was burning us up.

I had no water. But neither did the hyenas and so eventually they gave up and went in search of shade and water. Hans was right. I was eternally grateful. I watched them for a long time until they faded into the distance. Gradually my courage returned. Now, however, I had another problem. Time had passed and there was no longer enough time to reach Seronera before nightfall. My only option was to head for Simba kopjes where the main road to Arusha passed and hope I could find a vehicle to flag down. It was a Sunday and in those days there were few tourists so the chances were slim, but they were better than standing on that rock. So I walked again, keeping a wary eye out, and taking a circuitous route from rock to rock. Eventually I reached Big Simba Kopje and the road, and sat down to wait. My luck turned and a tourist vehicle came my way. They were happy to take me to Seronera; the thought of anyone being out on those plains alone seemed beyond credulity to them.

I stopped at Hans Kruuk's house. Staying with him was my Oxford supervisor, Niko Tinbergen.[23] I told them what had happened, to which Hans made the laconic comment that the hyenas were only curious. Curious about what, I reflected—how I tasted? Niko drew a cartoon of it in Hans's visitor's book. The story appeared later in Michael Crichton's novel *Jurassic Park*.[24]

* * *

Murray Watson followed on from the early work of the Grzimeks and Talbots to document in detail the migration pattern of wildebeest and zebra during the period 1962–6. He showed for the first time that the northern Serengeti was essential as the dry season refuge for the migrants, and that these migrants moved into Kenya. It was this information that led to the Masai Mara Reserve being set up in Kenya and the Lamai wedge being added to the Serengeti National Park in 1966.

George Schaller conducted the first studies of lions in 1966–9.[25] He demonstrated that lions depended on the migrant wildebeest for their survival; when the migrants were in their territories they produced cubs and had plenty of food. After the wildebeest left for the north the lions found food was scarce and cubs starved. Lions could not follow the wildebeest and had to remain within their territories because the cubs were too small to walk far and needed shelter and protection.

Hans Kruuk documented the ecology of spotted hyenas for the first time; he demonstrated that they were true predators and not merely the scavengers that popular mythology supposed. In fact hyenas are the most numerous predator in our ecosystem. They live in packs and commute from their dens to wherever the wildebeest are on the plains. After the migrants leave the plains for the woodlands he found that hyenas could travel as far as 50 miles to reach them, coming back to their dens every few days. However, hyenas could not reach the migrants once these went to the far north on the Mara River.

Hugh Lamprey, before becoming the Director of the Serengeti Research Institute, conducted one of the earliest studies on how different antelope species used different habitats and so could coexist with their different niches.[26] Both Peter Jarman and Richard Bell started with Lamprey's results and developed them further. Peter Jarman was one of the first scientists to show how the social behaviour of antelopes was determined by their ecology.[27] Large species could live in herds and escape predators by running. Small species had to escape predators by living singly and hiding. In turn, large species had to eat coarse grass because they needed abundant forage for their large bodies, whereas small species could pick the high quality buds and fruits because they needed only small amounts. So he showed how ecology and behaviour evolved together. Richard Bell took Vesey-Fitzgerald's idea of the grazing succession (see Chapter 9) and documented the sequence of grazers in western Serengeti, one of the best examples of how one species can facilitate the ecology of another.[28]

Meanwhile I was able to record from my observations of the tame buffalo what they liked to eat, which in Serengeti was purely grass. There were some grass species that they preferred over others, the species with soft stems and leaves that grew to medium height of 2–3. These species had high protein content and low fibre.[29] From these studies I was able to measure the amount of food available to buffalo (and later for wildebeest) and eventually show that they ran out of food in the dry season. But this is getting ahead of the story.

The counts of buffalo showed over the 1960s and early 1970s that numbers had continued to increase, births were numerous, and deaths were relatively infrequent. The animals had more than enough food and the population did well. What, therefore, had caused the population to drop to such a low level in the first place?

Veterinarians had been collecting blood samples from wildebeest for several years in the mid-1960s and I supplied them with further samples in the late-1960s and with samples from buffalo. The veterinarians found high levels of antibodies to the disease rinderpest in some of the samples, but not all of them. Rinderpest is a viral disease, and it gave rise a few thousand years ago to the human disease measles. Those animals with the antibodies had at sometime in the past been infected but had recovered. They were now immune because of the antibodies. Those animals without the antibodies had not been exposed. At the laboratory in Banagi I had the skulls of those animals from which blood had been collected, and I was able to assess their age by counting the layers of dentine and cementum in their teeth—each year a new layer is laid down. I discovered that in both species the animals that had the antibodies had been born before 1963 (wildebeest) or 1964 (buffalo). After those dates no animals had been challenged with the disease; it had simply disappeared. It was after 1964 that both species had shown the rapid increase that scientists had noticed from their counts.[30]

It was the disappearance of a disease that had allowed the increase in buffalo and wildebeest populations—an unexpected result. But it raised new questions. What was rinderpest doing in Serengeti in the first place and why had it died out? To answer these questions we have to go back in time and understand the history of Serengeti.

4

The Great Pandemic

W E are becoming increasingly aware that the environment that we live in is changing. It is not just the climate but also soils, plants, and animal populations. Some species are invading from outside the system. Populations of native species are also changing. Scientists want to understand the causes of all of these changes both to manage human ecosystems and to conserve natural areas. So when we found that the huge herds of wildebeest and buffalo in Serengeti were increasing we wanted to know why. To find the answer we had to look back in time, not just to the recent past but also into prehistory, because events during those ancient times set the scene for the present: they determine the geology, geography, and evolution of our present environment, and provide the clue as to why changes are now taking place.

We will see in this chapter how historical events provide the answers to the questions I asked at the end of the last chapter. We can also deduce by some detective work what happened in past centuries leading up to the events of modern times. By great good luck the Serengeti is also the site of considerable palaeontological research, which has given us an understanding of what the environment and species were like millions of years ago and up to present times. Serengeti is a very special area, one of the great wonders of the modern natural world. It is one of the last remaining places that resemble how the world was before the ice ages, during what is called the Pleistocene—older than 10,000 years ago. During that period there were many migrations of large mammals all over the world but most are now gone, the remaining ones highly modified. However, some are less disturbed than others and Serengeti may be one of those.

Ancient Serengeti

The Olduvai River runs across the Serengeti plains from Lake Lagarja in the south-east until it peters out in a temporary swamp called the Olbalbal depression. Over time this river has cut down through layers of deposits on the plains and created the Olduvai Gorge, and in this gorge scientists have found the evidence for what the Serengeti was like as far back as 4 million years ago. Other sites on the slopes of the Crater Highlands, particularly Lae-toli, have added to this information.[31] The Leakey family have been associated with these finds since the 1930s, making many famous discoveries.[32] More recent work is in the hands of a group from Rutgers University under Robert Blumenschine,[33] and it is this group that has shown how the ecology has changed.

Four million years ago there was dense savanna with some forest patches along rivers much like parts of northern Serengeti today. There were duikers and bushbuck species similar to those found in forest habitats today. By 3 million years ago conditions had become drier with a mixture of shrubs and small trees similar to far eastern Serengeti woodlands of today. Nevertheless there were lakes and swamps, because a giant buffalo-like animal lived in them. It had an ancestor called *Simatherium* but by 2 million years ago this had evolved into *Pelorovis*, twice the size of present-day buffalo and with horns 6 feet across turning downwards compared with the African buffalo's 4-foot upward-turning horns.[34] This was the lake that is now the temporary Olbalbal swamp.

These habitats persisted until 1 million years ago when more open savanna developed, becoming drier still. With the more open habitat, there were more grazing ungulates, the white rhino occurred then (it is now absent from Serengeti), and there was an absence of species such as reed-bucks and hippo, indicating that lakes and rivers were not present. Sometime between 500,000 and 100,000 years ago the present treeless plains developed.[35]

Hominids occurred throughout these 4 million years, gradually diverging into several species, some feeding on hard nuts, others developing scavenging abilities and turning to eating meat. Famously, Mary Leakey, who appears later in my story, discovered the footprints of one of our early relations in volcanic ash laid down about 3 million years ago.[36] In fact, human-like species have always been present in Serengeti. One fossil showed modern humans were present 17,000 years ago. The earliest modern humans were hunter-gatherers, much like the Hadzabi hunters who are still present at Lake Eyasi next to Serengeti today. With the abundant wildlife that seems to have been always present there were good opportunities for hunting.

In the past 20,000 years the environment has changed back and forth from wet to dry, and with these swings Lake Victoria has either expanded, covering what is now the Ndabaka floodplains, or contracted such that Speke Gulf would have disappeared and the Serengeti would have spread some 100 miles westward. In wetter times lowland forest spread around the southern and eastern shores of the lake and western Serengeti.[37] We see from these records that the Serengeti has been changing constantly both in environment and in species over several million years, and we should expect it to continue to do so.

Historical Serengeti

Early history in East Africa is fragmentary because there are no written records and what we know is gleaned from oral history and archaeological evidence.[38] Hunter-gatherers were the original inhabitants of East Africa and remnants, such as the Hadzabi near Lake Eyasi, are still present today. They would have hunted the Serengeti region over the past few thousand years. Early pastoralists were in Africa some 6,000 years ago as indicated by rock paintings in the Sahara—a region that was then savanna.[39] One thousand years ago Arab traders from Arabia were travelling along the coast, purchasing ivory and slaves in small numbers. They did not at that time travel inland and limited themselves to trading with Bantu peoples who lived along the coastal lowlands of East Africa. At about the same time Bantu peoples were also living inland in what is known as the interlake region, between the two branches of the rift valley. This was the area that included Uganda and Tanzania west of Lake Victoria to Lake Tanganyika. Between the inland and coastal Bantu peoples was a strip of plains country too dry for agriculture. Early pastoralists moved south from two groups, the Nilotics down the Nile and the Cushitics (Hamitics) from Somalia, displacing the hunter-gatherers. They essentially moved into unsettled regions during the past thousand years. The lake Nilotics, who include the Luo, settled around Lake Victoria in the eighteenth century, moving south and reaching the area west of Serengeti in the nineteenth century. Their distribution was limited by the presence of tsetse fly and it was not until the mid-twentieth century that they were able to spread east close to the present Serengeti National Park border.[40] Other Nilotic groups occupied the dry grasslands and plains along the rift valley but were limited by savanna to the west that contained tsetse fly; only the Serengeti plains were used by pastoralists. The Barabaig who currently live south of Lake Manyara represent these early pastoral peoples.

Sometime in the sixteenth century the interlake Bantu emigrated east into what is now Tanzania from the Congo forest. The first migration of the Wasukuma tribe is thought to have been in 1548, moving south around Lake Victoria and then spreading east and north. This was the largest tribe in Tanzania, but it was divided into some twenty sub-tribes or kingdoms. They were agriculturalists with an advanced ironworking industry.[41] These migrations of Bantu into East Africa occurred during the 'Little Ice Age' (1450–1850) when the northern hemisphere was cooler than today, which generally means the savanna of East Africa was drier. The agriculturalists spread along the east side of Lake Victoria but were held up by tsetse fly west of the Serengeti at least until the 1920s.[42]

In the 1800s the Waikoma, an offshoot of the Sonjo on the eastern rift valley, moved west to their present location north of the Serengeti corridor. Their main village was Ikoma, now called Robanda. They were the first settlers in this area and lived largely by hunting in the unsettled wildlife areas of Serengeti until the 1930s.[43] In response to attacks by Maasai shortly after they had settled they moved to encampments surrounded by wooden fences on the top of hills along the Grumeti River. The plains below, in what is now the western corridor, were used for hunting ungulates, much as today.[44] In the

1870s missionaries crossing the Serengeti from the east reported that agriculture was not found in the Serengeti woodlands until they reached the village of Nata. The boundary of agriculture is on the higher ground running east–west from Ikoma, through Nata to Bunda, leaving the silty and waterlogged low ground to the south for wildlife as it still is today.[45]

The eastern side of Serengeti is the domain of a much more recent incursion of pastoralists, the Maasai. They moved south from Sudan through the rift valley, reaching Ngong in Kenya by about 1640. They did not enter Tanzania until the nineteenth century, keeping to the open plains of the eastern rift valley and, because of their more warlike disposition, displacing Bantu and earlier Nilotics in their path.[46] They depended on cattle, so they were able to live only where there were no tsetse flies, which carry the trypanosomiasis disease, nagana, that kills cattle. Tsetse like dense savanna bush because their pupae require shade to mature in the soil. So the Maasai avoided the Serengeti savanna and were confined to the eastern plains. The Maasai arrived at Ngorongoro as recently as 1850, evicting the previous inhabitants.[47] Maasai used only the open treeless plains and even in the 1870s, not long after they arrived, they did not venture further west than Simba kopjes in the middle of the long grass plains; this is reported by missionaries, who produced a map of where people occurred in that period (see Map 3).[48] The first recorded crossing of Serengeti was in 1852 by an Indian, Juma bin Mbwana. He met the Maasai coming south in the rift valley, but then marched across the uninhabited plains and savanna of Serengeti to reach Mwanza. In the following decades explorers wanting to reach Lake Victoria went either south via Tabora or north towards Uganda from Mombasa to avoid the 120 miles of uninhabited Serengeti plains.[49]

The decade of the 1890s was a disaster for Maasai on the Serengeti plains, struck first by rinderpest, which resulted in famine and smallpox. The starving peoples emigrated and were captured by other tribes. The next decade saw civil war between three family groups, exacerbated through restrictions imposed by the Germans, at the end of which the population was almost wiped out. However, during the First World War, when the Germans were on the run, the Maasai began to re-establish their lost ground. Even in 1905 warriors ventured west, establishing temporary barracks at Moru Kopjes from

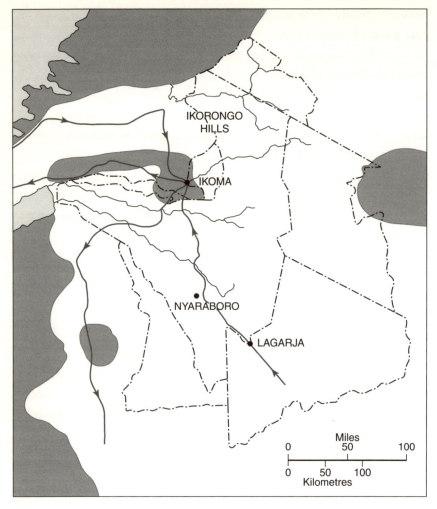

Map 3 Route of Oscar Baumann across Serengeti in 1891 showing the distribution of agricultural and pastoral peoples (shaded).

which to raid the Wasukuma further west. The Maasai never set up permanent camps west of Olduvai, even in the twentieth century.[50]

The best description of the Serengeti ecosystem before the great disturbances of rinderpest, which I describe below, comes from Oscar Baumann, who led the German anti-slavery expedition from the coast to Lake Victoria in 1891–3, passing through Ngorongoro and Serengeti.[51] He was a geographer interested mostly in the peoples and their culture, but he provides some description of vegetation and wildlife. In those days travel was by foot with porters and armed guards (askaris). Arriving from the east coast, and having crossed the Maasai plains south of Arusha, he discovered (for Europeans) Lake Manyara, then Ngorongoro Crater, and then Lake Eyasi. On 27 March 1891 he left Lake Eyasi and headed across the eastern Serengeti plains to Lake Lagarja (Map 2). Before reaching this lake he visited Maasai bomas where the inhabitants were starving. They had goats but no cattle, and they begged him to give them cattle.

This should have been the beginning of the rains but conditions were still very dry, with clouds of dust and no wildlife except long lines of ostrich. Lake Lagarja, which is a shallow alkaline lake, was almost dry but he describes flamingos along the shore. Leaving the Maasai behind at Lagarja he walked west across the plains, in the next few days coming across many skeletons of wildebeest that had died from the rinderpest, and reached the hills at Moru Kopjes.[52] From there the caravan headed north, through the hills until it reached the Orangi River and met Waikoma hunters from the nearby Ikoma village;[53] it was here they first came across agriculture. Until then they had been in wild acacia savanna without cultivation or people. After a few days in Ikoma he headed west, crossing the Grumeti River at a place called Nyasiro,[54] and headed towards Speke Gulf on Lake Victoria. The caravan passed many villages with cultivation on the high ground, and crossed uninhabited plains on the low ground with large numbers of wildebeest, rhinos, and antelope (probably topi). Baumann did not see any live buffalo, only skeletons of animals that had died of rinderpest. This area was the Sibora Plain, now in the Grumeti Game Reserve, and it has remained the same to the present. At the lake, he saw huge herds of wildebeest, antelope, and zebra. Conditions were still very dry and the migration was using the lake as a water source. Only near the papyrus-lined shore at a village called Katoto was there cultivation, scattered huts, and drying fish.

After travelling by boat south to Mwanza and back to the head of Speke Bay, he toured west and north for two months, eventually reaching the region near the Mara River some 20 miles west of the northern Serengeti. Here, he says, 'east of us was spread out uninhabited areas' to the edge of Maasailand. So the northern Serengeti was uninhabited in the late nineteenth century even before rinderpest.

Returning south he passed Ikoma again and crossed the western corridor at Musabi and the Dutwa plains; he saw big herds of antelope (topi) where they still occur today. Further south, near the Duma River, they met the cultivation of the Wasukuma, approximately where the boundary of the Serengeti National Park is today.

Baumann's journey is described here because this is the first, and only, description of the ecosystem before the effects of the Great Rinderpest changed the area in the following decade (see below). We see from his map, where he clearly demarcates cultivation, that the boundaries of the uninhabited areas are similar to those of present times if a bit further out (Map 3). The vegetation was more open parkland, suggesting large acacias with little regeneration. Wildebeest, zebra, and antelope were present, and rhino were everywhere. He makes the comment: 'there are astoundingly many rhinos. They have the luck of not having a valuable tooth and so can increase unmolested, whereas an extermination campaign is carried out against elephants, as...they are extinct.'[55] In today's context, with almost extinct rhino and expanding elephant numbers, the irony is breathtaking. Baumann saw no elephant in Serengeti, the after-effects of the ivory trade.

The ivory trade of the 1800s

Arab traders had for centuries been exporting ivory and slaves to Arabia and India, but on a relatively small scale. In the mid-nineteenth century, however, the Sultan of Oman moved to Zanzibar Island and made it his permanent home. It was from that time that trade with the interior of East Africa expanded rapidly. Both ivory and slave numbers exported from East Africa rose exponentially.[56] Captain Sulivan describes how the British Royal Navy was hunting slave dhows from 1849 onwards in a futile attempt to stop the trade and free slaves.[57] Zanzibar was the base from which Arab caravans set out for the interior of Africa. The main staging post was Tabora in the centre of what is now Tanzania. From Tabora these caravans went west into Congo and north along both the western and eastern sides of Lake Victoria. One such route went across the Serengeti plains, through the central woodlands, and then west to Lake Victoria.[58] At a place near what we now call Nyaraboro Plateau, close to Moru,[59] Farler comments there is a 'tribe of elephant-hunters, who neither cultivate nor keep cattle, but live entirely upon the flesh of the animals they kill in hunting. They supply the caravans with a great deal of ivory. Their country is full of elephants and other big game. They do not mix at all with other tribes.'[60] So elephants were abundant in Serengeti in the 1860s and 1870s. Other routes started from the Kenya and Mozambique coasts, and again from Khartoum travelling west and south. These routes effectively covered east, central, and southern Africa.

The purpose of the caravans was to obtain ivory, which was sent to India and Arabia. The ivory export trade had been in operation on a small scale for a thousand years or more, but starting around 1840 the trade expanded rapidly as the demand for ivory in Europe and North America suddenly increased. The fundamental reason for the increase in demand was the Industrial Revolution led by Britain in the 1780s. By 1840 the increase in wealth in Britain created a demand for luxuries such as pianos and snooker tables—ivory was used for piano keys, billiard balls, and even knife handles. Imports of ivory to Britain were steady before 1840 but thereafter increased linearly to about 1880.[61] Exports of ivory from Zanzibar and Khartoum dropped to near zero after 1890 as the elephant populations collapsed from over-hunting.

The Zanzibar Arab caravans purchased ivory from the African tribes who hunted elephants within their territories. The Arabs set up staging posts, the first of these in 1845 at Tabora in the centre of the country, and another in the 1860s at Ujiji on Lake Tanganyika. The traders supplied guns to local people who then systematically reduced elephant populations.[62] Local populations were plundered for porters who were forced to carry the ivory in chains to the coast. This was part of the slave trade, which reached its peak in the 1860s when some twenty thousand slaves a year were brought to Zanzibar. Half were kept there for the clove plantations, the rest re-exported along the East African coast.[63]

The hunting was so heavy that elephants were eradicated. This extirpation started near the east coast in the 1850s, and then extended westwards as the caravans were obliged to travel further to find ivory.[64] Tsavo Park, Kenya, now famous as an elephant park and in the early nineteenth century the domain of the Wakamba elephant hunters, had no elephants in the 1890s.[65] Similarly no elephants were observed in Serengeti during S. E. White's expedition of 1913 (except in the Masirori swamp on the lower Mara River[66]), nor were they seen in the 1920s as we shall see later.[67]

Recovery of the elephant population in East Africa took the first half of the twentieth century, starting from the few remaining groups in remote areas, for example in 1897 in the Lorian swamp and along the Turkwell River of northern Kenya, and the Nile in Uganda, and in 1904 in the Aberdare mountains of Kenya.[68] It is likely that the recovery was delayed because the original elephant habitats had been taken over for cultivation in the colonial period of the 1910s and 1920s.[69] However, by 1908 elephant herds had reappeared in Tsavo.[70] Tribal elders remembered seeing two elephants in the west of Serengeti in the 1930s, the first they had ever seen.[71]

It was not until the 1950s that elephants were seen commonly both in and out of protected areas. In the early 1950s Serengeti had less than 100 elephants, but by 1958 there were some 600 and in 1961 counts recorded over 1,000. By 1965 elephant numbers in Serengeti had increased to 2,000 and a decade later they had reached 3,000.[72] The dense vegetation that was a feature of savanna Africa in the first half of the twentieth century provided abundant food and so allowed elephant numbers to increase. The same increases were observed in most savanna areas of Africa where elephants lived. Although some of this increase must have been due to compression from expanding human activity outside, it was certainly not the only cause for increase or even the main cause; the populations were growing back from the low levels of a century earlier.[73]

The general conclusion we reach from this history is that the trends in elephant numbers and vegetation biomass observed in the 1960s can be seen as normal adjustments following the major disturbance to elephant populations a hundred years earlier. Understanding ecological history over the past two centuries has allowed a better interpretation of what happened in Serengeti and elsewhere, and this will help us with conservation practices for the future.

The great rinderpest pandemic

Rinderpest ranks as one of the greatest socio-ecological disasters in human history, on a par with bubonic plague in Europe in the 1350s and smallpox in the New World after 1492. It is a viral disease of cattle that occurs naturally in Asia. It was introduced to Ethiopia in 1887 by cattle brought from India by Italian invaders; there is no evidence that it was ever in Africa prior to this event. It spread to West Africa, then south through East Africa in 1890, Malawi in 1892, and reached the Cape by 1896. The resulting pandemic killed over 95 per cent of cattle throughout Africa; in many cases complete herds died off, and famines decimated the human population two to three years later in many parts of Africa. By 1892 widespread famine occurred in Ethiopia, Somalia, southern Sudan, and eastern Africa.[74] At least three-quarters of a million people died in Tanzania in the mid-1890s.[75] Clear evidence that rinderpest swept through Serengeti in 1890 comes from explorers' reports of starving pastoralists in 1891. Oscar Baumann, passing through Ngorongoro and later Serengeti, in March 1891, when confronted with starving people writes,

These people ate everything available; dead donkeys were a delicacy for them; but they also devoured the skins, bones, and even horns of cattle. I gave these unfortunate people as much food as I could, and the good natured porters shared their rations with them, but their hunger was unappeasable and they came in ever greater numbers. They were refugees from Serengeti where starvation had depopulated whole districts. They had fled to their countrymen who had barely enough to eat themselves. Swarms of vultures followed them, waiting for victims. We were daily confronted by this misery and could do almost nothing to help. Parents offered us their babies in exchange for a piece of meat. When we refused to barter they artfully hid the children in our camp and escaped. Soon our caravan was swarming with Masai babies and it was touching to see how the porters cared for the little ones. I employed some of the stronger men and women as cowherds and thus saved quite a number from death by starvation.[76]

The surveyor, G. E. Smith, who travelled through the northern Serengeti in 1904, referred to the Mara area, Kuka Hill, and east to the Rift Valley escarpment as formerly inhabited by Maasai but was then depopulated. He noted 'the Masai themselves attribute the cause of their decline...to the cattle plague (rinderpest), which devastated their country in 1888..., and to a bad attack of small-pox, which followed in the train of the famine caused by the wholesale death of their cattle'.[77]

The repercussions from this pandemic have had a profound influence on the ecology of the Serengeti over the past century. Buffalo were present in the region before the rinderpest, because the explorers Stanley and Emin noticed abundant tracks of this species when they travelled through this area in 1889. Then the epidemic hit. We have a detailed account from John Ford of the important ecological consequences of rinderpest in the region south-west of Serengeti.[78] Both Maasai pastoralists around Ngorongoro and Bantu agriculturalists west of the Serengeti ecosystem suffered severe declines from the rinderpest due to the resulting famine and secondary outbreaks of smallpox. Consequently there was an evacuation of large areas surrounding the ecosystem. Lack of cultivation by humans and lack of grazing allowed the bush to grow up, and this was colonized by tsetse flies as soon as the wild herds of antelopes had recovered; tsetse originated from the uninhabited Serengeti and spread westwards towards Lake Victoria.

It was not until the 1930s that the spread of the tsetse fly was stopped with the help of extensive burning regimes and mechanical bush-clearing methods. Julian Huxley, the famous evolutionist, was sent to Tanganyika by the British Government in 1929 to advise on ecology and conservation amongst other things, and around Mwanza he commented on both the migrations of zebra and gazelle and the policy of burning the thornbush, a policy that continued for several decades thereafter.[79] Eventually the reduction in dense savanna vegetation allowed the subsequent expansion of the human population in the 1950s, which has continued at an accelerating pace ever since.

Before the rinderpest epidemic, wildlife was abundant. The explorers Speke and Grant passed through the Maswa country, just south of Serengeti, in 1861, and described typical savanna and ungulates such as buffalo and zebra that are still found there today.[80] With the arrival of rinderpest, buffalo and many species of antelope, particularly wildebeest and hartebeest, were decimated. Baumann mentions as he walked across the Serengeti plains the skeletons of wildebeest and later those of buffalo.[81] F. C. Selous, the famous hunter and naturalist of the nineteeenth century, said of the buffalo:

48

Till the end of 1889 and the beginning of 1890 it was however exceedingly common being found all over the country where there was good grass and water....The real stronghold of the species was, however, the Masai country, where with perhaps the exception of Burchell's zebra and hartebeests, it was the most common of all the big game....on the Mau plateau they were also abundant and might be seen in dense black masses on the open grassy downs at all hours of the day.

In 1890 rinderpest appeared amongst the native cattle and spread among the buffaloes so rapidly that by the end of April they were decimated and there are now [1900] few left.[82]

Several authors have corroborated Selous's account of the effects of rinderpest on wild ungulates. In 1901 Hinde and Hinde wrote, 'Ten years ago the buffalo was the commonest animal in East Africa but owing to rinderpest and pleuro-pneumonia, it is now reduced to some three or four herds.'[83]

The buffalo population in East Africa remained at very low numbers between 1900 and 1920. We know this from reports by American President Theodore Roosevelt[84] and the Kenya game warden Blayney Percival.[85] John Patterson, who built the railway through Tsavo in 1898, also travelled through Kenya in 1907 and commented that on one occasion he met a buffalo and a rhino simultaneously, and set out to hunt the latter, which he regretted because he saw fewer buffalo than rhino.[86] In August 1913 the hunter S. E. White walked through what is now the northern part of the Serengeti, and although he refers to seeing a few buffalo, they were very scarce compared with present times (see Chapter 5). In the 1930s, Audrey Moore,[87] wife of the first Serengeti game warden, Monty Moore, had to make a special journey to northern Serengeti in order to see them. Buffalo remained uncommon until the 1960s.

* * *

Rinderpest continued to affect the Serengeti buffalo and wildebeest until the 1960s. Epidemics occurred every ten to twenty years over the period 1900 to 1963.[88] The virus then disappeared from wildlife populations as a result of a cattle vaccination campaign. The evidence for this disappearance came from antibody titres that virologists measured in blood sera collected from animals during the 1960s. We were able to calculate the age of these animals from their teeth and found that nearly all (80–100 per cent) of the wildebeest and buffalo born in the 1950s had rinderpest antibodies, but only 50 per cent of those born during 1960–2 had antibodies; and none born after 1963 suffered from the disease. So the disease had affected the animals during the 1950s; it died out in the early 1960s and was gone after 1964. By eliminating the disease from the domestic reservoir, the vaccination programme effectively protected wildlife from infectious yearling cattle, and consequently the disease died out rapidly in wild populations.[89]

Only ruminants, those that regurgitate and chew their food,[90] are affected by rinderpest, and the species most affected are those most closely related to cattle. So buffalo were the most affected followed by wildebeest and then other antelopes. Infections were reported in giraffe and warthog but other ruminants appear to have been less affected by the disease.

A consequence of the disappearance of rinderpest in 1963 was the doubling of survival in wildebeest and buffalo calves, and both populations increased rapidly. We shall see in later chapters the further impacts of these huge changes in numbers. However, one important clue that rinderpest was the fundamental reason behind the great changes we were to see in Serengeti came from zebra. They are non-ruminants and so are not affected by rinderpest. Their numbers have remained constant at around 200,000 throughout the forty-five years from 1958 to 2003. This implied that the changes in the ruminant populations were due to something that affected only them, namely rinderpest.

The account of the rinderpest shows how the historical evidence from the 1890s combined with the antibody data over the period when it disappeared provided the answer to our question at the end of Chapter 3—how did rinderpest arrive in Serengeti and how did it disappear, allowing the rapidly increasing populations that scientists encountered when they first studied them in the 1960s? The effects of the rinderpest pandemic are still being felt some 120 years later. What happened in Serengeti, however, between the outbreak of the great epidemic in 1889 and the events that we recorded in the 1960s? To answer that question we must go back to the history of the colonial period, first under the Germans between 1890 and 1920, and then the British from 1920 to 1961.

5

The German Era

B Y remarkable luck the travels of Oscar Baumann in 1891 took place only
a few months after rinderpest had hit the Serengeti region; however, the
vegetation had not yet had time to respond to changes in grazing, fire, and
other influences. So Baumann's account shows us what the ecosystem looked
like immediately before the impact of the disease. What happened to the
ecology of this area following rinderpest and during the German era from
1891 to 1920? We have a few reports that provide clues to the changes that
were taking place. First the Anglo-German boundary commission established the border between British East Africa (Kenya) and German East Africa
(Tanzania) in 1904, which crossed the northern Serengeti. Then an American
hunter, Stewart Edward White, travelled through the Loita and Mara area of
Kenya in 1910–11 and the Serengeti in 1913.

The Germans had a strong influence on Tanganyika (then called German
East Africa), including the Serengeti, until the end of the First World War.
They had set up a chain of forts, of which Fort Ikoma on the edge of Serengeti
was one, most of them built by 1896.[91] These forts communicated via signal
posts on mountain tops using heliograph. At least one of these posts was on
the hills east of Serengeti, probably Kuka Hill, and another in the central
ranges west of Seronera. These communicated with both Fort Ikoma and
with settlements on the far west of Serengeti, at the edge of Lake Victoria on
the Belila Hills—where Bunda now stands. Ikoma was an old slave-trading
post along one of the Arab caravan routes but was taken over as a German
government fort.[92] In 1901 the German encampment at Ikoma was attacked

51

by Maasai raiders. This attack led the German authorities in 1902 to build a fortress on the hill in the shape of a square with watchtowers on either side. It was supplied by foot porters from Mwanza along a route that ran across the western corridor.[93] One German officer with a troop of African askaris (infantry) was stationed there. Lt Paul Deisner had to control the European gold prospectors who indulged in poaching with nets. He stayed until 1913. After the First World War began, the Germans sent an expedition north and captured the British town of Kisii. Stung by this event the British retaliated and captured Fort Ikoma in 1916. Throughout this period there was little activity in the Serengeti itself except for the gold mines. The Waikoma elders recount that although no one lived in Serengeti, they would use the area around Banagi Hill for hunting.[94]

The Germans had at least two routes to Arusha across Serengeti: one crossed the plains from Lake Lagarja and probably reached the Seronera River, then Banagi, before arriving at the three small gold mines called Nyabogati, Kilimafeza, and Konogo. From Kilimafeza the route went to Fort Ikoma—from the air one can still see the line of the road through the trees past the kopjes where the Bilila Hotel now stands (Konogo mine), and on to Fort Ikoma. The other route went directly across the northern Serengeti from Loliondo to Ikoma; it has now disappeared.[95]

In January 1904 G. E. Smith carried out the survey of the Anglo-German boundary from Lake Victoria to Kilimanjaro; he crossed the northern Serengeti and Mara Reserve.[96] Travelling east from the lake he mentions that for 20 miles west of the Isuria escarpment on the British side the landscape consisted of highland grassland with patches of forest, uninhabited except for wandering bands of Wandorobo hunters, and provided good grazing with abundant wildlife. This fits the scene at least until the 1980s. South of the border human settlement occurred further east than on the British side but was confined to the area above the Isuria escarpment. Below the escarpment (in the current Serengeti) there were no inhabitants. He describes the magnificent view of mountains in the distance. A photo of the Mara River at the boundary shows it to be much narrower than at present but with more substantial gallery forest. On either side of the river were wide plains with sand rivers. No people were met with between the Isuria escarpment and Kuka Hill.

The hunter S. E. White travelled in 1910 south from near Narok over the Loita plains (the very northern part of the Serengeti ecosystem), then into the montane forests of the Loita Hills and on to the Nguruman escarpment of the rift valley. Two years later he made the first recorded expedition on the west side of the rift valley across the northern Serengeti to Lake Victoria (Map 4). In July and August 1913 he and his guide, the professional hunter R. J. Cunninghame, and a dozen or so porters started in the Nguruman forests in the east and walked across northern Serengeti just south of the Mara River, meeting great wildebeest and zebra herds almost all the way to Lake Victoria at what is now the town of Musoma, then only eight months old.[97] His descriptions of geography, vegetation, and animals are so detailed that we can trace his track within a mile or so. White mentions that south of them, unseen, lay

the Serengetti [sic], a grass plain many days' journey across, with a lake in the middle, swarming with game and lions; the Ssale [Salai], a series of bench plateaus said to be stocked with black-maned lions beside other game; some big volcanoes... with forests and meadows and elephants in the craters.[98]

He comments frequently on the abundance of animals. While crossing the northern Serengeti (from Bologonja to Wogakuria) he describes the landscape:

In the bottom lands were compact black herds of wildebeest, grazing in close formation, like bison in a park, and around and between them small groups of topi and zebra—two or three, eight or a dozen—moving here and there, furnishing the life and grace to the picture of which the wildebeest were the dignity and the power. And every once in a while, at the edge of a thicket, my eye caught the bright sheen of impala, or in the middle distance the body stripes of gazelle, or close down in the grass the charming minature steinbuck or oribi.[99]

A year earlier, in July 1912, the famous Kenya game warden A. B. Percival camped on the Mara River near the present Kenya–Tanzania border.[100] He noted that wildebeest were very concentrated in the area due to the prevailing drought. Topi were also numerous then.

White had told Leslie Simpson about the abundance of lions on the Serengeti plains and the latter had already taken one safari there by 1914.[101] The First World War interrupted further trips, Simpson did not resume his

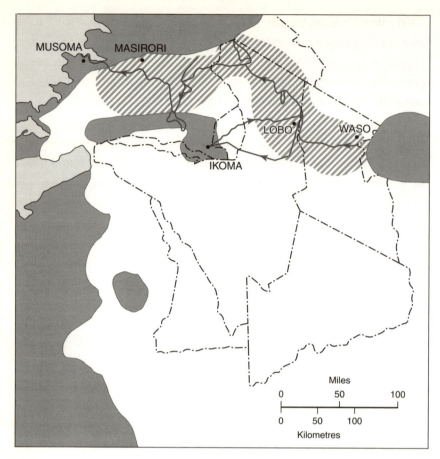

Map 4 Stewart Edward White's journey across northern Serengeti from Lake Natron to Musoma on Lake Victoria in the dry season of 1913. The distribution of wildebeest and zebra migrants (hatching) was then much further west and east of where they are now found in the 2010s. Distribution of people (shading) was similar to that in the 1890s (Map 3).

hunting safaris until 1920, and the Germans were left to themselves to fight the war.[102]

* * *

Following the rinderpest in 1890 young trees regenerated in the woodland areas so that the vegetation was relatively dense in the decade before the First

World War. This can be seen in photos from S. E. White's account of 1913. More importantly many areas west of the current protected area had dense vegetation and heavy tsetse fly infestations.[103] These prevented human occupation and allowed wildlife to live there. Baumann, Smith, and White all show that human occupation was limited to areas above the Isuria escarpment, west of the Ikorongo Hills far outside the present ecosystem boundaries, and north of the alluvial plains of the Ruwana River in the west (Map 2). In the east, Maasai were limited to areas east of the Kuka-to-Grumechen line of hills that form the present boundary. They were restricted by the presence of tsetse fly. Even the Wandorobo that Cunninghame met in 1913 had their base in the Ikorongo area west of the Park.

Large herds of wildebeest, zebra, and topi were recorded by White and Cunninghame in August and September 1913 from near Waso in the east almost continuously as far west as the Masirori swamp near Musoma. Clearly, these animals were part of the migration that would use lands along the Mara River at that time of year. Wildebeest, zebra, and topi were very abundant in 1913. Kongoni were abundant in the east. White noted all species that we expect to see today including steinbuck, dikdik, both gazelle species, impala, giraffe, and both species of reedbuck. Chanler's reedbuck, of course, was seen only on the highest hills in the east, where they are still found today. Rhino were everywhere and seen frequently, as were lions and leopard. Wild dogs and cheetah were also recorded across the north. These distributions are similar to those recorded at least until the early 1970s within the park boundary.

However, there are differences from the distributions of today. White's records are the first clear evidence that wildlife lived far to the west of the park in the absence of human settlement. Dense vegetation and tsetse fly contributed to this situation, which lasted until the 1950s; humans were not in this area when the northern extension was added in the late 1950s (see Chapter 6).

Roan antelope were clearly far more widespread; they were seen at Lobo, Bologonja, and west to the Ikorongo hills. I saw a small group at Lobo in January 1968 but otherwise they were confined to the Lamai and Mara triangle and the *Terminalia* woodlands of the north-west by the 1960s. They went extinct in northern Serengeti by the early 1990s. Waterbuck were more abundant, occurring in large groups of 50, which we do not see today. Buffalo were recorded in 1913 across the north but only as lone males or very small herds; in the 1970s they occurred in herds of thousands. Elephant were unknown to the German officer and local peoples at Ikoma and they were only recorded at the Masirori swamp on the Mara River near Musoma. This swamp would have been a dry season refuge for elephants, a refuge that is no longer available for them. Ostrich were more abundant on the plains, judging from Baumann's report. Greater kudu are indicated on Map 2 at Kuka Hill. In 1911 White shot a male on the Nguruman escarpment and so they were probably in the Loita forests. Kuka Hill had montane forest at that time and may have supported greater kudu. In general we see a far more widespread distribution of the great migration in the 1910s commensurate with a lack of human settlement. More importantly Baumann had reported that the migrants were using Speke Gulf for water during the drought of 1891. White's records confirm that the last refuges for the migration during severe drought were Lake Victoria and the Masirori swamp. Both water sources are now out of reach of the migrants.

Rarer ungulates such as rhino, roan antelope, waterbuck, and lesser kudu were more widespread and abundant and I will return to this aspect in Chapter 20. In contrast, elephant were very rare, consistent with the after-effects of the ivory trade of the period 1840–90 (Chapter 4).

With the end of the First World War the next major change in the Serengeti took place: the beginning of the British administration in 1920 brought with it an unrestricted slaughter of wildlife from foreign hunting safaris. They were to have profound and momentous consequences for the future conservation of Serengeti.

6

The Beginning of Serengeti

T HE Serengeti plains were known, as we have seen, for their abundance
of lions as far back as the 1910s, the area being a favourite for foreign
hunting expeditions. Stewart Edward White's hunting trip in 1913 has already
been described in Chapter 5. The First World War interrupted sport hunting
for a few years but by 1919 hunting safaris had commenced[104] after Tanganyika was mandated to the care of Britain by the League of Nations.

In 1914 Leslie Simpson pioneered the route for vehicles from Nairobi via
Narok south through the northern Serengeti to what is now Seronera. After
the war he set up a camp on a spring at the south end of Kuka Hill, initially
known as Simpson's Camp but later taken over by Al Klein and becoming
Klein's Camp by 1926. This route has remained essentially unchanged to this
day as the road from Nairobi to the Maasai Mara Reserve and south to Serengeti. This road arrived at the northern entrance to the park, the gate near
the spring being called Klein's Camp in 1950. In the 1990s the gate, with the
same name, was moved to its present position a few miles east. In the photographs taken by Martin and Osa Johnson in the wet season of 1926, Klein's
Camp spring is surrounded by dense hillside, *Croton* thickets, and fig trees,
but all of these have now disappeared from the frequent grass fires of the past
fifty years. Leslie Simpson set up his camp on the edge of the plains at the
kopjes where the Seronera lodge was built later in the 1960s. This became the
new Simpson's Camp,[105] only much later changing to Seronera after the river
that runs past the kopjes.

The photographer-hunters Martin and Osa Johnson first went to the Serengeti in 1922. Martin was an adventurer who signed up as a cook on Jack London's voyage to the New Hebrides in 1907, although he did not know how to cook. How he satisfied the crew is a mystery but luck went his way when he learned to use both still and movie cameras from some French photographers. Then Jack London fell ill and the expedition was left to him to bring home. Writing a book of these adventures brought him fame and his wife, Osa, whom he married in 1910. Together they set out on a life of adventure in Africa. On their first trip in 1922 they indulged in hunting, as everyone did at the time, but then they changed to filming, Martin cranking the camera while Osa guarded him with the rifle—all of 4 feet 11 inches, she was a crack shot. They paid a short visit to the Mara plains where they observed large herds of wildlife including elephant, rhino, lion, and cheetah, but no further details are mentioned.[106]

Their second expedition, 1924–7, was more substantial but most of it was spent in Kenya. However, in July 1926 they drove to the Serengeti for about six weeks with George Eastman, of Kodak fame, and Carl Akeley, the famous collector for the American Museum of Natural History. They stopped at Klein's Camp then on 23 July moved on to the Mbalageti River near Lake Magadi. There they filmed the wildebeest migration, commenting on tens of thousands of them. They also photographed topi, giraffe, and many lions around the edge of the plains and a few miles down the Mbalageti valley. Giraffe herds were exceptionally large, up to eighty in one group. The vegetation was dense mature *Commiphora* trees and young regenerating umbrella acacias, very different from the same area in the 1960s when it was open acacia parkland similar to that in the 1890s described by Baumann.[107] Eastman recorded a small buffalo herd 10 miles downstream.[108] Having shot one and then a rhino, which were very common, he commented that he now had all the 'big five' except elephant for which he had to go to Kenya—indicating the absence of elephant at that time. As Eastman departed, heading north to Kenya, they saw a pack of 12 wild dogs, confirming the presence of dogs in the woodlands.

The third Johnson expedition in 1928 lasted three months in Serengeti from May to July.[109] Again they camped on their way south at Klein's Camp, then at Banagi Hill, and finally set up a base at Seronera amongst the kopjes. In the following weeks they travelled across the plains to Simba kopjes, camped at Moru kopjes, and in June drove down the Mbalageti valley to the Ndoho plains and the Duma River looking for lions to film.[110] There on the Ndoho plains they met 'huge herds of zebra, topi, wildebeest and others' as well as lions and hyenas of course. They also saw 'storks; tens of thousands of them, both the common marabou and a smaller variety that we had never seen before'.[111] These numbers of marabou are far greater than we have seen in modern times. The other species was likely to be Abdim's stork, an intra-African migrant that even now occurs in thousands at the beginning of the rains. Back at Moru they filmed the migration moving through Moru, noticing how the large herds of zebra preceded the wildebeest. They also met Wandorobo, nomadic hunters who were following the migration. At Seronera they estimated 100,000 antelope, including all the species still present years later in the first decades of the twenty-first century, and numerous lion, hyena, and cheetah. In both 1926 and 1928 the migration of wildebeest off the plains and into the woodlands occurred in July, later than in any year that we have observed since 1960 when movements into the woodlands have generally been in May and at the latest by the end of June. No one at that time had any idea where these migrating animals came from or where they went. Indeed the Johnsons speculated that they came from the Athi plains near Nairobi. There were several other hunting safaris at that time including the Lieurance expedition of November 1928–March 1929.[112]

In general, what we learn from those expeditions is the greater abundance of topi and giraffe than today, and that the migrants travelled, then as now, from the plains at the end of the rains into the Mbalageti and Seronera valleys. These animals returned in large numbers, especially zebra, in the opposite direction, from the woodlands onto the plains, with the heavy rain of December 1928. There was not yet a recognition that these movements were part of a migration. These expeditions recorded the presence of wild dogs and roan antelope through the northern areas of Serengeti. Also, there was no one living in the areas where they camped. However, the most important information comes from their photographs of vegetation, which have proved invaluable by showing how the habitats have changed since then—I have been able to locate where many of these photos were taken using well-known hills and kopjes as clues, and repeated the photographs at intervals since 1982. We shall return to this in Chapter 11.

Their fourth expedition in 1933 was by far the most ambitious.[113] For this they shipped two Sikorsky amphibious planes to Cape Town, South Africa, and then with pilots flew them north 4,000 miles to Nairobi. Sometime in July they flew to the Maasai Mara and then Serengeti, constructing airstrips at Keekerok and Seronera, both of which are still in use; in those days nothing was there. They camped again at Seronera and Moru for some three weeks only. They filmed the wildebeest herds from the air between Naabi Hill and Gol kopjes, and a rough estimate of 90,000 can be seen. As in earlier years it is significant that the migrants were still on the plains in July. However, this was only a part of the migration for the Johnsons flew their plane to the Ndoho plains some 30 miles west, camping on the Mbalageti River and filming on the Duma River. They set up a hide, baited the lions with a topi they had killed, and filmed Osa watching from the back of the plane. On one occasion a lion started to climb into the plane looking for something inside—Osa had to hit it on the head with a bag of flour. They mentioned large herds of wildebeest, zebra, and topi.[114] The wildebeest population, therefore, would have been over 100,000 animals at that time. This number matches what we would expect of a herd recovering slowly from the devastation of the great rinderpest epidemic forty years earlier but still suffering from the disease on an annual basis.

After eighteen months Osa became ill and so they flew back along the Nile to Cairo, then across the Mediterranean to England—they had flown the whole length of Africa. Although they intended further expeditions, their plans came to a sad end when Martin Johnson was killed in a plane crash at Burbank, California in 1937.[115]

* * *

Meanwhile professional hunters were discovering and using the northern Serengeti and Mara. Denys Finch-Hatton saw and photographed over 100 lions at Klein's Camp in 1925,[116] and he escorted the Prince of Wales to the eastern Mara plains during the royal visit to East Africa in 1928.

Syd Downey first saw the area west of the Mara River in Kenya in 1927 and kept it as his secret hunting domain for several years until he teamed up with Donald Ker in the 1930s. Downey hid his crossing points on the Mara River so that others could not follow him. This area, which later became known as the Mara Triangle, was teaming with wildlife, and the Ker and Downey hunting firm continued to use the area until the 1950s.[117] This pair[118] were responsible for the development of hunting controls in the 1940s after the Second World War, and together, with Mervyn Cowie, pushed for the first national parks in Kenya.[119]

Throughout the 1920s the Seronera area of Serengeti was the playground for what Finch-Hatton called 'tourist hunters'. Simpson led one group that included S. E. White in 1925, which killed 50 lions. Other parties followed, some accounting for as many as 100 lions on one trip.[120] Lack of concern and supervision by the Tanganyika Game Department led to unethical practices, such as shooting from vehicles and excessive killing. One American group killed some thirty lions in this way in 1927. Another massacre of 80 lions took place on the Mara plains in 1928.[121] Such wanton slaughter so infuriated Finch-Hatton—later immortalized as the lover of Karen Blixon in the film *Out of Africa*—that he wrote a letter in 1928 to *The Times of London* denouncing both these indiscriminate practices and the irresponsible Tanganyika administration. The colonial government replied by denying they had occurred, but in subsequent letters Finch-Hatton published the facts and substantiated his accusations. He pushed for some form of protection. Questions were asked in the British parliament and the Tanganyika administration was ordered by the British Government to do something about it.[122]

The British Government also sent Julian Huxley on a fact-finding tour of East Africa in 1929, and his report recommended that the Serengeti plains be set aside as a national park (along with other areas in Kenya and Uganda).[123] Finch-Hatton and Huxley had been at school together and knew each other well, and although there is no direct evidence it is likely they influenced each other in these issues. In any event the Tanganyika Government declared a small area from Banagi Hill westwards along the corridor between the Grumeti and Mbalageti Rivers a closed reserve in 1929 and dispatched a game warden to Serengeti in 1930. Denys Finch-Hatton, who must be credited with the creation of Serengeti as a wildlife conservation area, was killed in his single-engine Gypsy Moth aircraft on 14 May 1931.

* * *

The posting of Major Monty Moore[124] as the first game warden signalled the beginning of Serengeti as a protected area for conservation and the end of the free-for-all massacres of lions. He built a bungalow in 1930 facing the north side of Banagi Hill overlooking the Mugungu River, a house made from mud bricks, and plastered with whitewash. It was the headquarters of western Serengeti until Seronera was developed in 1959. The house was subsequently taken over by scientists and used until 1973—it was our home for three years in the late 1960s.

In 1930 a larger 'Serengeti Closed Reserve' was demarcated, which included the present Serengeti, the Ngorongoro conservation area, and the Loliondo district as far north as the Kenya border and east to the edge of the Gregory rift valley (though none of these names were used at that time).[125] The 'closed reserve' in the 1930s provided nominal control of hunting, though in practice the wardens had little ability to supervise, and abuses of hunting etiquette continued. Nevertheless, Moore worked at getting protection for lions and in 1937 a 900-square-mile area was declared the 'Serengeti Game Sanctuary'.[126] The boundaries were between the Mbalageti and Grumeti rivers as far as Speke Gulf in the west—what we now call the 'corridor'—and along the old road to Naabi in the east from Banagi to about Lake Magadi.[127] In 1937 the colonial government announced its intention to create a national park and in 1940 the Serengeti National Park was formally declared.[128] However, no action was taken due to distractions of the war until 1951. In that year the boundaries were finally agreed after much discussion and the park proclaimed on 1 June 1951.[129] The boundaries of the park established in 1951 included the western corridor to the Mwanza–Musoma road, east along the Orangi River to the Ngorongoro Highlands. The southern border followed the present NCA to the western edge of the plains then north to Seronera. The park headquarters was based at Ngorongoro with a western outpost situated at Banagi. In the early 1950s it became clear that the park, as it was with people living in it, was becoming untenable, and so the Tanganyika Government decided to excise those areas where the Maasai had traditionally lived, leaving the remainder as a national park without people—the area that had never had residents except for nomadic hunters as far as records existed back into the mid-1800s (see Map 3, Chapter 4).[130] A committee of enquiry was set up to make the decision.[131]

The problem was where to draw the boundaries. To decide these boundaries one needed information on both the location of pastoralists and the movements of the migrating wildlife herds about which almost nothing was known. A commission of enquiry in 1956 employed Professor Pearsall to make the study.[132] He reported that there were two groups of migrating wildebeest, one from the Ngorongoro Crater that moved onto the eastern Serengeti plains in the wet season, and one from the corridor that used the western plains. In between the two there was a gap not used by either, a gap that could conveniently be used for the boundary of the new Serengeti National Park. This advice was accepted and the new boundary running north–south across the middle of the plains was legislated in 1959 (Map 2). The information that Pearsall used was more hearsay than fact and was incorrect—the Ngorongoro wildebeest never went onto the eastern plains but remained on the highlands or in the Crater itself. Instead the main migratory wildebeest of Serengeti used the whole of the plains including the Salai plains at the far eastern edge of the Gregory rift valley. Since the new boundary alignment the migratory wildebeest have moved outside the Serengeti National Park each year during the wet season. These movements have caused a confrontation with the Maasai because wildebeest carry malignant catarrh fever, a disease that kills cattle. Maasai herds move away from the plains when the wildebeest are there.

In 1958 Bernard and Michael Grzimek were invited by the Director of National Parks, P. G. Molloy, to conduct the first wildlife survey of the Serengeti ecosystem. Part of their work documented the movements of the migrants and they showed up the fallacy in Pearsall's report. However, by this time it was too late and the boundaries had been settled despite strong protests from the Grzimeks.[133] To compensate the Serengeti National Park for the loss of the eastern plains the authorities changed the western and northern boundaries. The south-western boundary, originally running along the edge of the plains, was moved westwards to include Moru Kopjes and the Nyaraboro and Itonjo hills. Moru was the only area that was then used by about a hundred Wandorobo hunters on an intermittent basis. The Maasai did not live there but used the area for dry season grazing. At the request of the Maasai elders the administrators moved the Wandorobo elsewhere— they were the only people to be moved on the plains.

In 1954 there were 194 Maasai who lived in the Gol Mountains but used the western plains on a seasonal basis.[134] Since the mid-1930s Maasai from the Gol Mountains had developed a seasonal grazing pattern: as the wildebeest left the western plains in June or July the Maasai followed them to the edge of the plains along the Ngare Nanyuki River in the north and towards Moru in the west to make use of spring water. They used these areas for some three months until the wildebeest returned in November, whereupon they retreated back to the Gol Mountains. This ebb and flow was constrained by ecological factors, namely tsetse fly, which prevented their further movement into the savanna, and the malignant catarrh fever carried by wildebeest, which caused the Maasai to move east again at the start of the rains. After 1959 their grazing routine was changed to make use of the Olduvai and the upper Ngare Nany-uki River water sources. The new park boundaries were designed to suit the other sections of the Maasai living in the Gol Mountains and at Olduvai.[135]

The northern extension—the whole area north of the Orangi River to the Mara River—was added (Map 2) because this area had never had inhabitants due to the presence of tsetse fly;[136] no people were moved out. It was a remark-able stroke of luck, perhaps the greatest piece of luck in conservation history, because this area turned out to be the essential dry season refuge for the migration, indeed the most important area of the whole ecosystem—without it the migration would have collapsed and the Serengeti reduced to just another sample of savanna with resident animals. Bernard Grzimek's distri-bution maps showed the animals moving north but far to the west and he opposed this addition because it appeared unused by the migration—his maps were simply too inaccurate and he misplaced the migration routes.

When the NCA was created they decided to fence the western entrance of the Angata Kiti valley in the Gol Mountains so as to exclude the wildebeest and keep the area for cattle.[137] When the wildebeest and zebra migration arrived on the plains in 1960 they went straight into the fence, several hundred thousand at once, and the fence fell over. Sensibly they did not try to put the fence up again. The Serengeti park headquarters, originally at Ngorongoro, were moved to Seronera where they remained until 1998 when they moved again to Fort Ikoma, outside the park. Both Chief Park Warden Gordon Harvey and deputy warden Myles Turner moved to Seronera in 1959. The Maasai Mara Reserve in Kenya was formed in the late 1950s under Maasai administration from Narok. The Lamai Wedge in Tanzania, between the Mara River and Kenya border, was added in 1966, thus creating a continuous protected corridor for the wildebeest migration from the Serengeti plains in the south to the Loita plains in the north. A small area north of the Grumeti River in the corridor was added in 1967. Subsequently the Grumeti and Ikorongo Game reserves were established in 1993, and the Wildlife Management Areas were decreed in 2003.

* * *

Sandy Field became Chief Park Warden in 1963. An ex-Provincial Commissioner from Sudan, he was a gentle, likeable eccentric. He gave the impression of a bumbling academic when in fact he was astute, perceptive, and clever. One Christmas he contracted the local tailor in Seronera to make him a suit of sackcloth, and this he wore at Christmas and other festive occasions, leaving the local inhabitants quite nonplussed as they watched him cycling about the village in it. Steve Stevenson took over for a short while in 1971 until David Babu became CPW in 1973, a post he retained through the difficult years of the economy until 1983 (see Chapter 8). Babu became Director of National Parks in 1985 and stayed until 1993. Meanwhile after four more CPWs,[138] Justin Hando took over from 1998 to 2006 (Chapter 17).

Myles Turner was instrumental in developing the field force into an efficient unit during his years as anti-poaching warden (1956–72). A wiry, weather-beaten bushman, an expert hunter, he turned to conservation with a passion in 1956. He helped the Grzimeks with their 1958–9 aerial surveys and so was introduced to flying, which he subsequently took up, becoming a most careful and reliable pilot. He was enthusiastic about research after the Grzimeks's visit and helped record the first accurate information on the movements of wildebeest and of burning patterns in the early 1960s, but after falling out with Murray Watson[139] he had little time for scientists and gave them no credit.[140]

The Serengeti National Park was one of the first areas to be proposed as a World Heritage site by UNESCO at the Stockholm conference of 1972 and it was formally gazetted as such and as a Biosphere Reserve, along with the Ngorongoro Conservation Area, in 1981.

With the final designation of park boundaries the stage was set for scientists to start their work in earnest. Apart from the Grzimeks, who had already conducted their first censuses in 1959, Lee Talbot also conducted an initial study for the International Union for Conservation and Natural Resources in 1960.[141] This proved valuable for Lee documented the wildebeest population when rinderpest was still very active and so gave us a baseline with which to compare after rinderpest disappeared. It was an important start.

7

The Migration of Birds

E VERY year millions of birds arrive in the savannas and grasslands of Africa from Europe and Asia for the winter. The problem is that most ecosystems in the world are full; the populations of species are generally at the limit of their food supplies and this is particularly so for the tropics. This raises the question of how these millions of birds can fit into such a small area, savanna Africa being much smaller than all of Europe and Asia combined. It implies that Africa is largely empty of birds to allow these northern migrants, in contrast to our understanding of tropical ecology. Reg Moreau, one of the original ornithologists living in Tanganyika in the period 1930–50, pointed out this conundrum as far back as 1950.[142]

The question is of special concern for conservation because migrating species require three areas that must be protected rather than the normal single area for most other species that do not migrate. First, migrants need a refuge area during the worst times of year when resources are scarce. For the birds that travel between the northern arctic or temperate regions and the tropics the refuge is in the tropics during the northern winter. Second, they migrate because they can make use of very abundant food during the northern summer and so increase the number of babies they can produce (compared with those that stayed in the tropics); and these breeding areas are, in most cases, far away in northern latitudes. However, these summer food supplies are temporary so the birds must leave again when the food disappears in autumn. Third, they need a corridor along which to migrate between the other two areas. Many birds can overfly inhospitable habitats such as oceans, deserts, or human modified lands, but even in these cases they need to land at staging areas on the way in order to rest and sometimes stock up on food. Shorebirds often use estuaries, mudflats, and beaches for this purpose. Mammal migrants of course need a complete and uninterrupted corridor between breeding and refuge areas. Africa, therefore, was a refuge for northern birds. But how did they fit into tropical savanna ecosystems when they are supposed to be full of resident species?

* * *

It was in 1963 that Reg Moreau suggested to Arthur Cain,[143] a professor at Oxford University, that this question would be an interesting research problem. Arthur agreed to let me tackle this project while he visited his graduate students in Africa in July–September 1965. He had two PhD students. One was Richard Bell, who was studying grazers in western Serengeti, and the other was Peter Jarman (see Chapter 3), who was at Lake Kariba in then Southern Rhodesia.

Reg Moreau was retired in Oxford and I was able to meet with him and learn about African birds. I went to Nairobi at the end of June 1965. There I met the Chief Park Warden of Serengeti, Sandy Field (Chapter 6), and with his driver, Onyango, we drove west to Narok, south across the Loita plains, and into the Mara and Serengeti on 1 July. I spent the first month on my own with some introduction by the resident scientist, Murray Watson, who was documenting for the first time the great migrations of the wildebeest. I lived at Banagi in a small wooden prefab down by the river and had use of a somewhat dilapidated pickup Land Rover. With this I set about recording the birds and their habitats in Serengeti since we had to know this first to establish how the migrants could fit in.

* * *

August 1965. First experiences of life in the bush can be dangerous simply from lack of knowledge but one learns how to cope, and a robust sense of humour is a help. I was visiting Richard Bell at his camp in western Serengeti on the Grumeti River. Apart from wanting to learn about his work I also needed to record the birds that were occurring on the western floodplains of Serengeti and in the dense riverine forests that occur along the river. I asked Richard if he knew of any good patches of forest. He enthusiastically said he would show me a patch a little way up river. Together we drove to this patch and very quietly walked in along a narrow path through very dense undergrowth of bushes and shrubs and a tall canopy of trees. It was extremely thick and we could see only a few feet in front and to the side. We were very quiet because we did not want to disturb the birds and there were plenty of them around, particularly black-headed gonoleks, white-browed robin-chats, and Schalow's turacos. I was in front, Richard close behind me. Suddenly a bush immediately in front of me erupted with violent shaking and a deep roar emanated from its midst. Instinctively I leapt to my right without waiting to see what monster was attacking and ran as fast as I could. After a few seconds I heard Richard give a frantic yell; I could hear thumping immediately behind me and something grabbed my ankles. I leapt for my life into the air and grabbed a liana. It was but the work of a moment to haul myself some 12 feet into the air. Relief was momentary because with a sickening crack the liana snapped and I found myself on the ground again. The landing, however, was relatively soft for I had fallen on Richard. I asked him what it was that had chased us but he was just as mystified. I stood up and gazed back at the way we had come while he rubbed his side ruefully. Our tracks seemed remarkably undisturbed for a large beast to have followed us. It was then I noticed that Richard was gazing myopically in the wrong direction—he had lost his glasses. 'Where are your glasses?', I asked him. He explained that he had heard this roar and seeing me disappearing into the bushes assumed I knew what it was and so decided to follow me. However, he had tripped over a log and fallen head first, hitting my ankles and losing his glasses. Peering myopically skywards he saw the liana swaying in front of him and had decided to grab it. Next thing he knew was my landing on top of him.

Carefully we searched back along our track and by good luck found his glasses not far away. Richard, now able to see again, joined me as we went back to the track since this was still the best way through the forest and we were not at all happy with our situation, not knowing what this animal was. We examined the track for spoor and were less than reassured to find tracks of elephant, rhino, and buffalo. We eliminated elephant because they make a trumpeting noise when charging, rhino snort and huff, which left buffalo and lion, both of which make similar deep-throated growls when charging. However, there were no claw marks of a charging lion.

As we were debating this issue and wondering whether to go forwards or backwards I noticed a pattering on the leaves around us and something wet landed on my head. Putting my hand up I found that it was a small piece of animal dung. Looking up I saw the branches high up thrashing about and then came a deep roaring call. Richard and I laughed so much we had to sit down as relief spread over us; we had frightened ourselves silly over a black-and-white colobus monkey. These colobus live in troupes of about ten animals and are territorial along the river, announcing their presence by roaring calls much like the howler monkeys of the American tropics. Usually living in trees they sometimes feed on the ground when it is safe. Because we had been so quiet we had caught one by surprise who leapt in fright into the bush in front of us, making a deep alarm roar.

Greatly relieved we made our way out of the forest. It was a long time before we told anyone, especially Myles Turner, the anti-poaching warden, who was sceptical that we were capable of surviving in the bush.

* * *

Arthur Cain arrived in September. By this time I had learned most of the common birds and found my way around a large part of the 5,000-square-mile Serengeti Park. My task now was to show Arthur the different parts of the park, the habitats, and the birds. We travelled from our base at Banagi north-east to Klein's Camp, where there were some dense thickets; Arthur wanted to see what lived in those thickets. We went in slowly, cautiously; it was a dangerous place to be, given the number of buffalo about. Bird watching requires spending much time looking up in the tops of bushes and trees, which is not the best thing to do in dense thicket when we should be looking at what is on the ground. At one point a buffalo erupted from within a bush where it had been sitting peacefully in the deep shade. When it had noticed our presence we were too close for it to escape, which is the circumstance that causes buffalo to charge. And charge it did. Our vehicle was some 100 yards away out in the open and I sprinted back to the vehicle. Arthur, who was in his mid-40s, did not proceed as fast. I looked back and had the vision of a somewhat puce figure, his legs going as fast as they could, and framed behind by a pair of huge horns; the buffalo was nearly upon him and he had the look of despair.

Realizing that some action was needed I ran to the side so that I was on the buffalo's flank. The buffalo, immediately sensing danger from the side, turned towards me, thereby letting Arthur get to the car while I had plenty of space to do likewise. I fairly hoisted him into the back of the pickup as I dived into the cab and the buffalo veered away having achieved his objective of seeing us off. Arthur was not overly amused by my cavalier treatment of this situation and I had to mollify him as we made our way back to Banagi.

In the end, during that visit I was able to show Arthur all parts of the eco-system and most of the important resident bird species that would have pos-sible competition from relatives migrating in.

* * *

Old male buffalo were a constant problem for us not only while out record-ing birds, but also at the small settlement by the Mugungu River at Banagi—a mile down the slope from the big house on the hill. There were three of us in wooden prefabricated houses, the entrance to which was a stable door that opened outwards above steps—it was a cumbersome entrance: one had to back down the steps to open the top and bottom halves. The grass was kept short around the houses.

My neighbours, each in a house of his own, were Fritz Walther and Jon Wyman. Fritz was a grizzled, wiry German, very excitable, and passionate about his work, which was on gazelles—he was a world expert on the behav-iour of the many species of gazelle that live in Africa and Arabia.[144] He tended to be a bit absent-minded. Our little group of houses had a generator for light-ing, and this was situated some 50 yards away. It was turned on at dusk and the last person up had to walk out to the generator shed to turn it off, thus walking back to the house in the dark. Because we were by the river, and we had short green grass, our settlement was a favourite feeding ground for several old male buffalo.

One night Fritz went out to turn off the generator, forgetting to take with him his flashlight. On his way back to his house in the pitch dark he heard a loud bellow just behind him and realized he was being charged by a buffalo. He raced for his house, the buffalo almost on him. There was simply no time, however, to get into his house through that cumbersome door, so he continued to run round his house, the buffalo hot on his heels. The only reason the buffalo could not catch Fritz was because he could not corner as tightly as Fritz could. But Fritz was tiring rapidly and could not keep this up much longer. He started shouting for help. Jon and I were both woken up by a terrified shrieking. We went out, shone a flashlight, and saw the startling scene of Fritz running round and round his house followed by a huge roaring buffalo. Half-dressed I raced for my pickup truck and drove straight at the buffalo, who finally veered off, blinded by my headlights. Fritz collapsed in a heap, exhausted and shaking, and it was several minutes before we could get him into his house again and dose him with whiskey. After that Jon and I found that for some reason we were the ones up last.

* * *

Buffalo were not of course the only source of concern, and life in the bush is not for everyone. Jon Wyman was an American student enrolled at the University of Nairobi in Kenya and studying black-backed jackals for his thesis. From time to time he had to camp at sites where he could view his animals at their dens. He took with him an assistant to look after his camp, an assistant who was not at all bush-savvy. In fact he was terrified. He was particularly convinced that lions were going to winkle him out of his tent and have him for supper. Despite much reassurance by Jon that no such thing would happen, he remained a reluctant camper. He insisted that when the two of them put up their tents, the two tents should be as close as possible, in fact so close that the guy-ropes overlapped each other. There was but 3 feet between them.

That night lions stampeded a herd of zebra on the plain near Banagi where they were encamped. The zebra took off in full flight towards the tents, most of them going around the tents, but a few went right between the two. The lion was close behind, grunting as it chased at full speed. The guy-ropes were instantly pulled up by the zebras' legs, there being no space between the ropes, and the tents promptly collapsed onto the faces of the two inhabitants—just as the lion raced through, stepping on the fabric with a deep rasping growl as it passed. The assistant realized that his worst fears had come to be—the lion had taken down his tent and was now about to extract him for dinner. He lay petrified, dumbstruck with fear, as the tent was raised up, the zip ripped open—and Jon poked his head in to ask how he was. The assistant finally recovered enough to say he was first going to sleep in the vehicle, and next day he was going home— for good.

* * *

By the end of the dry season we had recorded the general ecology of many of the resident bird species, particularly those that had close relatives migrating into Africa from Asia and Europe. The following months saw the arrival of the migrant birds and it soon became clear that they only appeared when rain brought large numbers of insects, many of them migrants themselves. This was the first clue as to how they could fit into an otherwise full ecosystem. We continued working on this problem long after 1965. It took another thirteen years to accumulate the information that was eventually published.[145] The migrants, it turns out, could not fit into the African landscape until after the rains started. As we noted earlier, the African seasons are determined by where the sun is relative to the equator. In the northern summer the sun is north of the equator and this brings rain to Sudan in July and August. As the sun moves south so does the rain, some six weeks after the sun. Eventually in the southern summer the rain falls in Zambia during November–February, six months out of phase with the north.

Birds arrive from the north first in Sudan (and other parts of Africa of the same latitude) where there is rain. Rain brings large outbreaks of insects. As the rain moves south there is a band of insects either hatching or even migrating, and this band follows the rain. For example, locusts, other grasshoppers, and some moths migrate with the rain. It is these insects that the migrating birds are following once they arrive in Africa. Many of these rainstorms are unpredictable, in both where and when they occur. So these birds are nomadic, following storms. In essence the migrants are avoiding competition with the resident birds by only going where there is a surplus of food. The answer to the question on how such a vast number of migrants fits into Africa turns out to be that they can only do so when the rains bring a surplus of insect food.

8

Socialism and War—Sort of!

THE first coordinated research into how the Serengeti ecosystem works
began with the Serengeti Research Project in 1962.[146] The main task
involved the detailed mapping of the great migrations of wildebeest, zebra,
and Thomson's gazelle. Work on predators began a few years later and the
whole group of scientists became incorporated into the new Serengeti
Research Institute in 1966 when a laboratory and houses were built near
Seronera.[147] The Institute was under the Tanzania National Parks but func-
tioned independently with a Director. President Nyerere, founder of the
nation of Tanzania, valued the nation's wildlife (I return to his contribution
in Chapter 20); he opened the Institute and was introduced to the scientists in
1973. Some of the research during this period has already been described in
Chapter 3 and I continue the story here. The Research Institute continued
until the breakdown in relations between Tanzania and Kenya in 1977
(Chapter 9). At that time the Institute was effectively disbanded and taken
over by the Serengeti National Park administration; there was no further
coordination of research and projects operated independently.

These changes to the research structure were a reflection of the gradual
development of the political background of Tanzania: to understand what
happened in the Serengeti during the years 1967 to 1986 one must go right
back to the start of independence in 1961 and follow the political evolution of
the country.[148]

* * *

75

On 9 December 1961 Julius Nyerere was sworn in as Prime Minister of Tanganyika by Chief Justice Ralph Windham. After one month he resigned in order to lead his party[149] and write his papers on philosophy and policy. Nyerere had been educated in Scotland and was strongly influenced by their ultra-left-wing trade unions. He became a world figure, leading the African continent in socialist principles, and a force behind the non-aligned countries during the cold war era. A man of high principle, his main focus was on improving the lot of the poorest, and largest, segment of society, the peasant farmers. But principle and practice were not the same, for his policies were turned to the advantage of the bureaucrats who had become the ruling class, while the peasants saw only decline in their living standards.

At first, political initiatives were slow and modest, distracted in 1964 by military rebellions and the merging with Zanzibar to form Tanzania. Development required capital investment from abroad but with such a poor capital infrastructure to start with investment was meagre at best. Furthermore, Tanzania had in 1965 major ideological conflicts with its main financial donors, namely with West Germany (over recognition of East Germany), the United States (over forged documents purporting to show US involvement in inciting rebellion), and Great Britain (over the Rhodesian Unilateral Declaration of Independence). All three withdrew financial aid and Tanzania was then forced to obtain aid from the communist bloc—East Germany, North Korea, and China. Faced with the lack of outside investment the government decided it had to take the risk and fund capital development itself.

On 5 February 1967 Nyerere made his famous speech, now known as the 'Arusha Declaration'. In this he laid out his philosophy of *Ujamaa*, or Socialism.[150] The slogan was 'Socialism and Self-reliance'. Banks were nationalized immediately as were the main import–export firms, insurance companies, and many of the major producing companies such as milling and tobacco. In due course farms owned by expatriates were taken over at nominal prices and they became government farms. Those who owned private houses had them taken over: clearly, capital assets were banned and income from such assets prevented. There was to be no private ownership of land, or of businesses, and foreign investment was banned. Instead there were to be cooperatives run by the government.

The most controversial policy was that of villagization.[151] From 1973 to 1975 a large majority of peasants were moved into small collectives of 250–300 families, with the intention that they worked for the group to provide exports. By 1977 some 13 million people lived in such villages, representing the vast majority of the population.[152] Ostensibly voluntary, the moves were in practice compulsory. These forced moves were started in the south of the country, which was far less organized and established, and were generally more popular, at first, than in the established north, where the moves were deeply resented.

Along with the nationalization of the banks the currency was changed. Pre-independence East Africa was bound together economically through the East African Community with a common currency as the European Union now operates. Tanzania began to undo this arrangement, starting with its own currency, the Tanzanian Shilling. Uganda and Kenya were then obliged to follow suit. Officially the three national currencies were held at par, but traders began to see them as different as Tanzania's and Uganda's economies began to decline relative to that of Kenya in the mid-1970s. In the early 1970s most of the major industries were nationalized and replaced by government corporations, which promptly made a loss and the government had to subsidize them instead of the reverse.

The price of food, fuel, and other essential items was set unrealistically low to benefit the peasant farmers who formed the majority of the electorate. Shopkeepers soon found that they could sell the same items at four times the official price on the black market—now the real value of the currency—keep the account books in order, and pocket the difference. By 1977 there were almost no items for sale on the shelves, and fuel had become intermittent in petrol stations. At the same time agricultural production stagnated; export of the main cash crops remained the same from 1968 to 1978.[153] Tourists were not encouraged and foreign income from tourism remained negligible. The lack of income from the government-owned farms, cooperatives, tourism, and the tax base meant that the government could not pay salaries for its by-now huge workforce. So it printed money, which sent the currency into an inflationary spiral.

The Serengeti was affected in a number of ways. Tourist hotels were nationalized and immediately began a decline in service. One privately owned hotel was at the historic old German Fort Ikoma, an ideal site for tourism as it overlooked the Grumeti valley with its migrating herds of wildebeest in the June to October period. This was taken over for the use of the Tanzanian senior army officers in May 1977.

Bureaucrats had gradually become the ruling class, taking over the powerbase from democratically elected members of parliament. Indeed, parliament was reduced to an institution with little relevance.[154] Government-appointed officials within the villages were more powerful than the elected officials, as was observed in a dispute involving the boundaries of Serengeti. In 1972 the Wakuria tribe in north-west Serengeti claimed the whole of the National Park area north of the Mara River for themselves. A meeting was held between the Serengeti wardens and the tribal elders, presided over by the powerful head of the TANU party, who was from that area. He summarily gave away the northern Serengeti and the tribe marched in with their cattle. They started hunting overtly. This resulted in an outcry in the conservation community, Professor Grzimek talked with President Nyerere, and the decision was reversed.[155]

Conservationists were becoming alarmed at the whimsical nature of decisions overriding legislation that had been passed by parliament. Where was the future of conservation under this ad hoc policy process? As we entered the 1970s we were to find out only too soon.

* * *

Meanwhile in the Serengeti, despite the political and logistical difficulties that began to affect the way we could do our research, we started to focus on how animal populations were limited and how species coexisted. Scientists in the 1960s were asking questions about competition between animals on two fronts. First, right from the time of Charles Darwin it had been proposed that the many different species that coexisted in an ecosystem did so by dividing up the resources in their environment so as to reduce the level of competition between the species—they called this interspecific competition. By resources we mean such things as food supply, and space for territories or nesting sites. The important implication of this theory is that resources must be in short supply or there would be no competition. The theory also predicted that closely related species should show differences in their ecology—where they lived, what they required for food, and how they escaped predators. The combined set of features that describes the place of a species in the world is called its 'niche'. The species, therefore, were partitioning their environment, or in other words showed 'niche partitioning'.[156]

On the second front, scientists had argued that if resources were in short supply, as implied above, then there should be competition between individuals of the same species—this is called intraspecific competition. If this was the case then populations should run out of food (or other resources) and stop increasing. This theory predicted that a population would be regulated, that is to say kept at the same number, by competition for a set amount of food. The theory predicted that as a population increased we should see an increasing proportion of animals dying from starvation.[157]

The East African savanna is an ideal place to look at niche partitioning for there are many closely related species of large herbivores living in similar surroundings. They are easy to study. We can watch them feed and avoid predators, and we can collect their dung from which we get other information on their nutrition. Were these species dividing up their environment? The first intensive studies in the 1960s focused on this question.

Large species are also ideal for a study of what is killing them, their mortality. How many die? What is the cause of death? Carcases of buffalo can be dissected and autopsied. Even hyenas cannot destroy a buffalo carcase. Wildebeest carcases are dismembered by hyenas but usually enough remains to give us the clues we need on their cause of death. The increasing numbers of both these species, which we had now discovered was due to the removal of rinderpest (Chapters 3 and 4), provided the opportunity to look for intraspecific competition. Were there increasing proportions dying from starvation?

The downside of using large species to study this question is that it takes a long time. In fact several decades are required to obtain a definitive answer, but at least we can find some clues after a few years as to whether populations are levelling out and what causes this. I set out to answer these questions using the Serengeti populations of buffalo and wildebeest—it took thirteen years.

* * *

Elephant numbers were also increasing and the same questions we were asking about buffalo and wildebeest also applied to them. Why were they increasing? What would stop the increase? However, there was a difference because elephant numbers were increasing all over eastern and southern Africa, not just in the Serengeti. It was because of this continent-wide change that elephant populations in Africa became a source of controversy.

Many of the early ecologists in the 1960s concluded that such increasing populations of elephants were unnatural; the theories of population regulation did not apply to them. It was argued that elephants lived such a long time—some fifty years and more—that they responded too slowly when they ran out of food. These scientists proposed that elephants stopped breeding rather than starved to death, a population response that was very slow and resulted in there being too many elephants for the habitat. Hence, the habitat was destroyed. Scientists pointed to the disappearance of trees in Uganda, Zimbabwe, and South Africa as evidence for their theory. These scientists concluded that elephant populations needed controlling and to do that they had to be shot in large numbers.[158]

Not everyone accepted these conclusions, however. Many conservationists, passionate about elephants, did not agree with this assessment. In East Africa the most vociferous of these conservationists was David Sheldrick, Chief Park Warden of Tsavo National Park in Kenya. He had been instrumental in setting up this park in the early 1950s just to save the elephants, and he was not about to see them being shot out again. He resisted the scientists' recommendations and, indeed, had them removed from his park. The controversy raged throughout the 1960s while numbers of elephants in Tsavo continued to increase.

What was still missing, however, was firm evidence that they did not die of starvation and responded only by reducing their birth rate. Then in 1971 a drought hit the area and a large number of elephants died.[159] This event conveniently removed the need for a culling operation, but had no noticeable impact on the polemic. Changes in habitat from the feeding of elephants were deemed to be unnatural and calls for culling continued.

It was not clear just how many animals were left in Tsavo. Clearly a count was needed, and so in September of 1972 the Serengeti scientists who could fly were asked to bring their planes to Tsavo to help with a census of elephants. Colin Pennycuick, Mike Norton-Griffiths, and I each took a plane, our complete fleet. The census in Tsavo following the drought of 1971 showed that half of the elephants had died; they had died because they had run out of food within reach of water, which was in the main rivers. The important consequence of this mortality was that the trees and bushes that had been removed by elephants then started to grow back and provide food for the remaining animals. The population was showing what might be called a 'downward correction', to borrow a popular phrase from economics. It meant that culling had not been needed and the population was on course for self-adjustment. The population was showing responses similar to those we were seeing in buffalo. Elephants were not different after all.

* * *

In January 1971 Idi Amin toppled Milton Obote as President of Uganda and assumed power. In September 1972 a force of a thousand Ugandan exiles, transported in Tanzanian National Service trucks and supplied with Tanzanian army rations, invaded Uganda across the Kagera River, which formed the border between Tanzania and Uganda. They bumped into resistance from Ugandan troops, and when after less than a day their ammunition ran out they fled back to Tanzania. Needless to say, Amin considered this an act of war.[160]

Amin retaliated by sending in his air force to bomb Tanzania, which is to say that he sent a twin-engine small plane across Lake Victoria to Mwanza. The crew threw grenades out of the window—they were lucky not to have blown themselves up—and then flew back to Uganda. It had the desired effect. Large numbers of townsfolk evacuated, thinking they were going to be bombed to pieces, camped out in the open, and then when nothing more happened marched back into town again. Chaos ensued for a few days. The Tanzanian Government reacted by imposing a flying ban on all small aircraft in the north of the country, which included Serengeti, of course. Presumably this allowed the military to shoot at any small plane since only the enemy would be flying. The ban continued for six weeks.

The first we heard of the flying ban was when we all returned to Nairobi after the Tsavo count to refuel and continue on our way home. The Control

Tower at Wilson airport calmly told us we were not allowed to fly into Tanzania. We were stranded in Nairobi.

The six of us—for each plane had a pilot and observer—met at my brother Tim's house. He was based in Nairobi while building an airfield for the Kenya military in the north. Our problem was communication with Seronera because there were no phones in those days. There was only high-frequency (HF) radio communication by national parks staff with their headquarters in Arusha. So we had to phone Arusha headquarters and ask them to transmit radio messages to the research staff in Seronera. Needless to say this was not only laborious; the messages also became scrambled and confused. After a couple of days it was agreed that we would fly the Serengeti research planes to Keekerok, the headquarters of the Maasai Mara Reserve, and park them there until we knew more about the situation. Our families then drove north from Seronera, over the border, which was open at that time, and met us at Keekerok.

Back in Seronera we were faced with a problem. Much of the research depended on the use of these planes, which were now 80 miles or so away in a different country. In particular we had a number of radio collars on wildebeest—collars that transmitted a signal to tell us where they were. I followed them in the Supercub aircraft using an aerial attached to the wing strut, and a receiver that sat on my lap in the cockpit. Wildebeest move large distances looking for food and water but if I searched every few days I was normally able to find the collared animals based on their previous position. However, if I left it longer than that, finding them became difficult and expensive as I had to search a large area. So it was important that I searched at frequent intervals.

The ten or so animals were spread over a large area of the ecosystem but in September many of them were in the north at their dry season refuge. I realized that I could find at least some of them in the Mara Reserve. At intervals of three days I drove back to Keekerok, and searched within the Kenya part of the ecosystem. Naturally I could hear the signal from some of the animals inside the Tanzania area, so when it seemed I was out of sight of border posts I flew into the Serengeti part hoping that no one would try shooting at the plane. I kept low to reduce visibility and also to give ground observers less time, which of course made tracking more difficult.

PLATE 1 Wildebeest on the short grass plains in the wet season.

PLATE 2 Wildebeest massing at Seronera before moving into the woodlands in June.

PLATE 3
Ol Donyo Lengai, the remaining
active volcano of the Crater highlands
that created the plains.

PLATE 4
Short grass plains with kopjes.
Crater highlands in the background.

PLATE 5
Moru kopjes at the boundary of the
plains and southern hills.

PLATE 6
Nyaraswiga, the easternmost of
the Central hills near Seronera in
the savanna country.

PLATE 7 The umbrella tree *Acacia tortilis* is the classic tree of the savanna.

PLATE 8 Kopjes in the woodlands are special habitats for many animals.

PLATE 9 Riverine forest on the Grumeti River in flood along the western corridor.

PLATE 10 Banagi hill, site of the first Warden's station in 1930 and the first scientific lab in 1960's.

PLATE 11 Entering data accompanied by the baby buffalo, 1968.

PLATE 12 The buffalo followed Anna and I everywhere around our Banagi station.

PLATE 13   A leopard found a warm spot next to Anna outside the tent.

PLATE 14   We stalked the buffalo herd. A thousand heads came up to look at us.

PLATE 15 Ruins of Fort Ikoma, built by the Germans in 1902.

PLATE 16 Roan antelope were everywhere across northern Serengeti before the 1960's.

PLATE 17 Wild dogs were once found throughout the savanna; they disappeared in 1992.

PLATE 18 Male buffalo were a constant problem when we were observing birds.

By November the flying ban was lifted and we were able to return to normal operations without fear of being shot down. The information on the track of these animals showed their route in detail for the first time,[161] and much later it was to prove valuable in showing where the migration moved outside the protected areas.[162]

* * *

Flying took on another form in the early 1970s for it was the beginning of hot air ballooning. Alan Root, a well-known wildlife cinephotographer, had brought a balloon to Serengeti for his film on the wildebeest migration.[163] Mike Norton-Griffiths decided that a balloon might also be a useful survey tool and brought one too. They were all learning how to fly them, sometimes the hard way. Alan had already made one film of his adventures, first when the basket settled into Lake Naivasha and the hot air was not sufficient to lift it out again; and later when they broke a record by flying over Mt Kilimanjaro from the Kenya side and landed in Tanzania, for which he was promptly arrested for illegal entry.

An unusual phenomenon took place at Lake Lagarja in March 1973. The wildebeest migration was on the plains near the lake and the calving season was well under way—there were hundreds of thousands of newborn babies. For unknown reasons a large group of wildebeest had decided to walk across the lake instead of around it. Whereas the adults could wade through the water and mud the tiny newborn calves could not, the water was too deep for them. As the herd progressed through the lake the babies turned back, unknown to their mothers until they reached the other side. At that point pandemonium broke out as thousands of mothers started calling for their lost calves, walking back and forth across the lake. Meanwhile the calves had gathered on the near side, most were covered in mud but many had already drowned. Possibly because the strongly alkaline mud obscured the smell of the calves the mothers rarely seemed to identify their calves even when they made it back to the near side. After two days the babies died. It was a die-off of catastrophic proportions; we counted over three thousand dead.

We went to investigate these events, camping at the lake. Mike Norton-Griffiths brought his balloon so that we could survey the scene from above the lake and obtain a better view of the mortality. Early one morning he persuaded me to come with him in the balloon; I was reluctant since he was still a novice pilot. We took off after the usual huff-and-puff of blowing up the balloon. Unfortunately there was a strong east wind once we were above the trees, which is often the case on the plains. We soon crossed the lake and were off westwards across the plains at a fast pace. The plains stretched some 20 miles to the hills and their accompanying trees. Landing requires open areas and preferably no horizontal speed. Since we were going lickety-split Mike was not keen to land; he hoped the wind might drop. It did not and after half an hour we were approaching the trees so we had to land anyway.

The art of descent, I am told, is to fire the burner sufficiently strongly to slow the balloon down but not so much as to cause it to rise again, and, because there is a significant time delay for the hot air to take effect, to do it well ahead of landing. At the moment of landing the ripcord is pulled, the canopy is opened so as to spill the hot air, and one finds oneself gently on the ground. That art requires practice. In our case things somehow got out of sequence. We came down with no trouble but a bit fast. Mike reacted by burning furiously but too late so we hit the ground very hard and bounced up again, aided now by the hot air taking effect. We rose high up in the air. Collision with the ground had caused both of us to lose our feet, sprawling at the bottom of the basket, just at the moment Mike should have pulled the ripcord. Instead he lost hold of it, and had to struggle to his feet first before regaining it and heaving. 'How far up are we?', he called. I peered over the edge, commenting it was too far. We were at least 50 feet up now, and coming down rapidly because there was no air in the balloon. We crouched in the bottom of the basket. We hit hard, but worse we were also travelling sideways very fast. The basket tipped on its side and was dragged for 100 yards or more, acting like a huge scoop as we clung to the inside. It gathered a large quantity of wildebeest dung, earth, and grass as we went, and when finally we came to a stop we were completely buried in dung.

Mike in his usual manner made light of it. 'Just a normal outing,' he said cheerfully. Nevertheless, we were now 20 miles from camp, we had no water or supplies, and as far as we were aware no one else knew where we were. Just as we were contemplating our situation with some concern, over the horizon appeared Anna with our two little girls, Catherine and Alison, strapped in the front seat. We were both relieved and totally surprised.

Anna had seen us heading westwards and realized that unless she kept us in sight no one would know where we would land. She leapt into the Land Rover and drove as fast as she could, while Catherine, then only three years old, gave a running commentary on which direction the balloon was going since Anna had to watch the driving over this rough terrain. For 20 miles she drove, just keeping us in sight as she crested ridge after ridge, until we finally disappeared altogether. She did not know what to do but decided to travel in the direction Catherine had last seen us and there in front over a ridge we appeared, now so covered in dung that our faces were invisible. The little girls considered this hilarious, the best part of the outing.

Ballooning, we discovered, was not much use for research, but it became a standard tourist attraction in both Serengeti and Mara and remains so today.

* * *

After the fiasco of Tanzania's invasion of Uganda in 1972 relations between the two countries remained at a frosty level throughout the 1970s, and this was a contributing factor in the collapse of the East African Community in 1976 (Chapter 9). The murderous Amin had systematically been killing intellectuals, military, and anybody he thought a competitor, throughout these many years. The Ugandan economy continued to collapse as anarchy swept the country, the currency devalued at an alarming rate, and Amin had difficulty paying the troops who kept him in power. His usual recipe was to create an international incident to take their minds off things they should not be thinking about.

Towards the end of 1978 he sent his forces into Tanzania across the Kagera River and annexed a small area of land. Nyerere gave him an ultimatum to get out or be pushed out. Amin ignored the ultimatum. This time the Tanzanians were better prepared, and the Tanzanian army not only pushed Amin back to the border, they then decided to advance into Uganda, taking the border towns. The local people welcomed the invaders, and so encouraged they continued on to Kampala in April 1979. Amin was forced to flee the country, remaining in exile until his death several decades later.

It had cost the Tanzanian Government a huge amount, some $500 million that they could little afford, to rid the continent of one of the most blood-thirsty tyrants in African history. No other country came to help; the West did not contribute although they were asked.[164] By early 1980 Tanzania was running short of essential items such as petroleum, wheat, maize, rice, and imported products due to the lack of funds.

Many in the army returned from Uganda with captured weapons, and quickly found they had a ready market to sell them, and perhaps their own weapons too. Soon the country was flooded with AK-47s, and for the first time gangs of bandits started to roam the countryside, hijacking buses, shoot-ing up shops, offices, and police stations. The Wakuria, living on the north-west boundary of Serengeti and belligerent at the best of times, got their hands on many of these weapons. They knew a good opportunity when they saw one. They started to shoot up both the wildlife and the rangers. So began a period of murder and mayhem that was to last seventeen years—and dev-astate the Serengeti ecology in the process.

But this is getting ahead of the story, for Tanzania had had another con-frontation with Kenya in 1977 that had profound and lasting effects on con-servation for the next two decades.

* * *

Scientists have over the past forty years documented the ecology of the ungulates, finding as predicted that each species had its own particular niche. Essentially species separate ecologically in four different directions. First, they differ in the type of vegetation and plant species composition in their habitats. Secondly, they differ in the type of food they eat, especially whether it is from trees and shrubs (browse) or grass. Thirdly, different spe-cies prefer to eat different parts of plants such as buds, leaves, or stems; and fourthly they separate along a gradient of aridity.

For example, different large ungulates prefer different parts of the habitat range from closed forest, dense woodland of decreasing canopy cover, savanna with scattered trees, and grassland, to semi-desert plains.[165] So, bush-buck live in dense forest or thicket, kudu prefer thickets in woodland, impala like savanna, and oryx are adapted to arid plains. Other species separate according to the type of food and the height off the ground that the food is found[166]—Grant's gazelle eat only flowering plants (dicots) in the ground layer (up to 2 feet), while eland browse shrubs higher up (1–5 feet), and giraffe eat tree shoots (3–10 feet).

Within the gradient of habitats there is a group of large herbivores that are all open grassland grazers, these being zebra, wildebeest, topi, kongoni, and Thomson's gazelle. Richard Bell and others have shown that these species, together with African buffalo, graze at different levels along valley sides from ridge top to valley bottom, a gradient called the soil catena. Thomson's gazelle prefer ridge tops, sandy shallow soil, and short grass; wildebeest and topi prefer mid-catena medium-height grasses; and buffalo concentrate at the valley bottom where soils are silty and deep. Each of these species separates out according to the size and component of grass eaten—buffalo like the leaves of tall grass and avoid any grasses shorter than 4 inches. Wildebeest prefer the leaves of very short grass. Zebra, however, can eat the coarse stems of taller grass.[167]

Different foods result in different shapes and sizes of the skull, mouth, and incisor teeth which pluck the grass. Tall coarse grasses require a wider mouth and incisor teeth, which tend to go with larger body size.[168] Grass-eaters have broad muzzles and wide incisor teeth designed for non-selective feeding, requiring high rates of food intake. Browsers have narrow muzzles with large central incisors and small lateral incisors designed for selective feeding. Larger body size not only allows a lower quality diet but also a wider range of habitats, which means a broader niche.[169]

The most closely related antelopes in Serengeti, topi and kongoni, differ along a moisture gradient, with topi in wetter areas and kongoni in the more arid. Where they overlap in Serengeti kongoni feed in mature grass stands while topi select green leaf in young growing stands.[170] In general, the predictions of niche theory seemed to be born out. The implication is that at least sometime in the past there had been interspecific competition to produce these differences, and so resources must have been in short supply.

* * *

By the mid-1970s our population work in Serengeti was beginning to throw light on what was limiting large herbivore numbers, a fundamental question that was taxing scientists and conservationists alike.[171] In particular we had begun to understand what determined the numbers of African buffalo, and evidence was accumulating for what limited the migratory wildebeest population as well, though there was still much to be done. At this stage we had established that animals were dying largely from starvation rather than from predation. Only about a quarter of all deaths were due to lions, the only predator large enough to kill buffalo.

A buffalo carcase is so large that even hyenas cannot destroy its bones, and it was the bones that gave us the clues that we were looking for. As we have seen, when animals are short of food they start to use up their fat supplies, which they had stored in their bodies during times of plenty. Such times in temperate regions are usually during the summer, though in desert regions high summer is often stressful, and times of food abundance are often in spring and autumn. In the tropical savanna it is the rainy season, when grass grows rapidly, that provides the good food. At that time herbivores put on fat under the skin, around the body organs such as the heart, the intestines, and above all the kidneys, and in the bone marrow of the limbs. During times of food shortage—winter in higher latitudes and the dry season in the tropics—animals cannot find sufficient food to supply their needs and they must make up the shortfall by using their own fat. The stores are used in a particular order, with those around the body organs and skin being used first and the bone marrow being used last. So if we examine the bone marrow and see no fat, then we know the animal had nothing left; it was in fact starving. Luckily, the bone marrow is difficult for scavengers to get at. When an animal is killed by predators the first parts that are eaten are the body organs and flesh. Lions do not eat bones but hyenas do and in smaller herbivores, such as antelopes like impala, the whole carcase is crunched up, skull and all. Just the hooves and horns are left.

When an animal dies from causes other than predation, usually the first scavengers to arrive on the scene are the vultures. Vultures spot carcases as they soar high in the sky. When one of them sees a dead animal it folds its wings and plummets like a stone, a very characteristic sight that every other vulture notices and follows suit. It is also obvious to the large carnivores, lions and hyenas, who are both happy to get a free meal. These carnivores sprint in the direction of the falling vultures. Consequently for vultures there is little time to get a meal before the carnivores chase them off. The large lappet-faced vulture has the ability to tear the skin and open up the carcase, the several other species having to wait for them to do so. Once the carcase is open, however, the far more numerous Ruppels griffon and white-backed vultures pile in, literally, often climbing right inside the body, and the body organs disappear within minutes, if not seconds.

Consequently all the evidence on the fat condition of smaller antelopes disappears long before scientists find the remains. In the larger species such as wildebeest, and especially buffalo, the fat around the body organs and skin is consumed rapidly but the majority of the skeleton is left for us to examine, especially the limb bones. We can check the marrow fat to see whether the animal was starving—even if it was captured by predators.

So we were able to show three things. First, the majority of buffalo were dying in the dry season and starving. Secondly, the amount of food was insufficient for the population; animals were reducing their food supplies from their own grazing. Thirdly, the death rate from starvation was increasing as time went on and as the population numbers built up. If this trend continued, then the population numbers should level off as the increasing death rate should eventually equal the birth rate.[172]

However, death from starvation is not a simple matter. Animals did not die from simple lack of food—that would take weeks if not months. In fact what was happening was that as the animal lost fat its immune system became compromised so that it was less able to withstand attacks from viruses, bacteria, and larger parasites. As we conducted autopsies on the carcases we found a large number of parasites that were afflicting the animals. Overall we recorded some eighty species of parasites. It seems that they lived in the animal without causing too much trouble, but when the animal became weak, one or the other of them could take over, increase to high numbers, and kill the host. The species that took over did so more by luck, a lottery of who got in first.

The most important consequence of this process is that it shortened the time for animals to die, causing the buffalo population to be more sensitive to food shortages. In essence this was a fine-tuned system that had evolved over millions of years where the buffalo were adapted to their food supplies and the parasites to their hosts. In addition, and most importantly, the plants were adapted to this heavy grazing; they had mechanisms to protect themselves by sending their own food stores underground, so that they did not suffer from overgrazing. Since this work in Serengeti in the 1970s scientists have shown that food limitation is a general phenomenon in ungulates around the world.[173]

This, then, was the system that has developed for an animal population in its own natural habitat, self-regulating with processes that worked well if they were allowed to do so without interference. But such a conclusion had to be tested. If the components of the system—plants, herbivores, parasites— have evolved to a finely tuned interaction in their natural environment then we should expect the system to show opposite behaviour if the components were not in their natural environment, and had never met each other before. We should expect to see extremes in the responses, such as populations showing high mortality, extreme evidence of starvation, severe disease outbreaks, and wide swings in numbers (instead of the normal small changes in population) often accompanied by major damage to habitats.

A related species of buffalo, the Asian water buffalo, had been released in northern Australia in the 1840s and by the 1970s had spread across a large area.[174] It was an ideal situation to test this theory of regulation by food and disease. When an offer from the federal government of Australia arrived to study the ecology of Asian water buffalo in the Northern Territory Anna and I gladly accepted it. We began this research in early 1974 but it was interrupted when Darwin was destroyed in a hurricane on Christmas day 1974. Later in 1975 we moved to Vancouver, Canada, where I took a position at the University of British Columbia and resumed work in the Serengeti.

Meanwhile, the research on water buffalo in northern Australia was continued by other researchers once Darwin became habitable. The results of the work showed that buffalo went through a series of increases and crashes caused by overgrazing of their swamp grasslands so that nothing but dry dust remained in the dry season. They died from starvation. Such extreme fluctuations in numbers confirmed our ideas that normal populations in their natural environments are regulated by the limitation of food modified by the effects of natural parasites. In abnormal environments, as in the case of the water buffalo from Asia that were in Australia with few natural parasites and sedges unable to tolerate heavy grazing, the delicate regulatory mechanisms are missing, and populations experience extreme swings in numbers.[175]

9

Border Closure

WE had been putting together the chapters for a book on the Serengeti,[176] and it had suddenly hit us that all aspects of the ecosystem were being shaped by the increasing wildebeest population, released from the scourge of the rinderpest virus in 1963. It was dawning on us that instead of independent studies of disconnected components, the group of scientists in Serengeti had all in fact been looking at different consequences of the same phenomenon without realizing it: Hubert Braun and Sam McNaughton had been looking at the grazing impacts of wildebeest on the grass communities as the migration proceeded; various of us, including Peter Jarman, Richard Bell, and Patrick Duncan, had been looking at how wildebeest were influencing other ungulate populations; while George Schaller, Hans Kruuk, Brian Bertram, Jeanette Hanby, and David Bygott were studying the responses of predators to wildebeest. Mike Norton-Griffiths was measuring the impacts of fire, Harvey Croze the effects of elephants. All these aspects of the system were shaped by the burgeoning wildebeest numbers. The story was coming together like a jigsaw taking shape.

Interspecific competition looked like the process shaping the ecological niche of the many species of large herbivores living in savanna Africa as I have explained in Chapter 8. By evolving different niches the species can live together. Because there are several ways of dividing up the niche space many species have evolved in this environment, and tourists see the legacy of this process in the great diversity of large animals in Africa.

However, there is another evolutionary process that has contributed to this diversity. Species do not just compete. They can also help each other and we call such an association 'mutualism' or 'symbiosis'. A classic example creates the organism we call 'lichen', where an alga and a fungus have evolved to live together to form a new organism. Sometimes one species benefits while the other is not affected either positively or negatively. We call this 'commensalism' or 'facilitation'. For example, cattle egrets follow wildebeest to take advantage of the insects disturbed by the grazers (if the second species is negatively affected then this is classic predation or parasitism, which I will come to in a later chapter).

Facilitation was first recorded amongst African ungulates in 1960. Desmond Vesey-Fitzgerald suggested that facilitation occurred amongst large grazers in the swamp grasslands of Lake Rukwa in southern Tanzania. Swamp grasses grow very tall, 7 feet or more. Elephants had no trouble eating and trampling these, thus creating open patches where the flattened grass sprouted succulent shoots some 2 feet high. The open swards were the perfect niche for buffalo that grazed down the grass to a shorter height. In turn the grazed swards were used by waterbuck, a large antelope that liked wet areas. Vesey called this sequence where one species creates a niche for another the 'grazing succession'.[177]

* * *

Richard Bell in the 1960s looked to see whether Vesey's theory applied to Serengeti. He recorded a grazing succession amongst resident herbivores in western Serengeti.[178] In that area there occurred together buffalo, zebra, wildebeest, topi, and Thomson's gazelle. In the wet season they all grazed together on the tops of shallow ridges where the grass was short and nutritious. Downslope the grasses grew taller and more fibrous, reaching some 4 feet in valley bottoms near rivers. As the grassland dried out at the end of the rains the larger animals moved down the slope first. There was a sequence of buffalo, followed by zebra, then wildebeest and topi together, and finally the Thomson's gazelle. As the species moved down they each grazed the grasses to a lower height—they were providing a niche for following species. Vesey's discovery applied to Serengeti.

This discovery begged the question: was there grazing facilitation between the great migrating herds of zebra, wildebeest, and gazelle on the main Serengeti plains? Bell speculated that this may be one of the factors allowing the migration and hence one of the reasons that Serengeti was unique. There was some intriguing evidence in its favour. Sam McNaughton, our grassland ecologist, had discovered in 1976 that when wildebeest moved off the plains at the end of the rains their grazing stimulated the regrowth of grass, producing an additional amount of short nutritious leaves that Thomson's gazelle preferred. The gazelle followed where wildebeest had grazed, a behaviour predicted by the facilitation theory.[179]

The facilitation theory for the migration could be tested. Since the wildebeest were known to be increasing we could predict that the gazelle population should also be increasing because of the extra food that wildebeest were creating by their grazing. The opposite prediction prevailed if interspecific competition was occurring—gazelle numbers should be decreasing.

The previous count of wildebeest had been conducted by Mike and I in 1973.[180] Since then no further counts had taken place and we were all wondering by 1977 what had happened: had they continued to increase as they had over the past decade or had they levelled off? The answer was critical for three reasons. First, the conclusions on the regulation of herbivore numbers through food supply (Chapter 8) could be tested by changes in the wildebeest population. Secondly, the idea that facilitation could be the process allowing both the huge numbers and the underlying cause of migration—the very essence of the Serengeti ecosystem—predicted population changes of other species if the wildebeest had continued to increase. Thirdly, wildebeest appeared to be driving all other changes in the ecosystem and so changes in their numbers should be reflected in changes in other parts of the system. Ultimately the management of the Serengeti ecosystem depended on our knowledge of changes in wildebeest numbers. Mike Norton-Griffiths and I set out to count the wildebeest in April 1977.

* * *

The East African Community, set up prior to independence, was an economic union of the three East African countries, Kenya, Tanzania, and Uganda.

They shared the East African Airways, the East African Railways and Harbours, a joint tax system, and a common external tariff. They preferentially traded goods and services between each other. Kenya, however, always had an advantage for it had been more developed industrially to start with. Multinationals invested in Kenya in preference to the other two countries, thus maintaining that advantage.

In 1970 Tanzania accepted an interest-free loan from China to build the railway between Tanzania and Zambia. About half the local costs of the railway were paid for by importing and selling Chinese consumer goods and using the proceeds to pay for the local costs. The Tanzanian government State Trading Company purchased large amounts at the Canton Fair in China at very cheap prices. However, these imports resulted in Tanzania severely reducing its imports from Kenya, and what was imported had to be at competitively cheap prices, contrary to the spirit of the East African Community. Naturally the Kenyans felt excluded from their markets in Tanzania, and after a number of years resorted to obstruction. They held on to railway wagons, prevented ferries on Lake Victoria from departing, and seized aircraft of the East African Airways.[181]

The ultra-socialist Tanzania, which had felt increasing antipathy towards capitalist Kenya over the past half-decade, was unlikely to accept these manoeuvres meekly. Instead, they schemed and plotted. In April 1977, without warning, Tanzania seized all Kenyan tour vehicles and light aircraft inside Tanzania, about fifty in all. They also closed the border and prohibited any aircraft that had landed in Kenya from continuing to Tanzania; from now on all international flights had to originate from elsewhere.

The hapless tourists, waking up that fateful day in Serengeti and expecting to be taken on a game drive, instead found themselves effectively captives of the Tanzanian immigration authorities. They were unceremoniously bundled into the captured vehicles, taken the 200 miles to Arusha, and dumped into hotels. Next day they were herded into one of the hotel dining halls and harangued by the minister responsible for tourism, Mr Ole Saibull.[182] They were lectured on the wrongdoings of Kenya, and the unfairness of the tourism trade that benefitted Kenya but not Tanzania.[183] Warming to his theme, Mr Saibull threw in the evils of capitalism for good measure. Then the tourists were forced to pay for a chartered plane to fly them back to Nairobi.

That, lectured Mr Saibull, should teach everyone a lesson. From now on tourists would have to outfit their safaris, and start and end their tows in Tanzania. Since Tanzania had by far the best wildlife areas tourists would obviously come to Tanzania, which should now capture the majority of the tourist trade, and for good measure poke a finger in the eye of Kenya. Tourism had always been a sore subject for the Tanzanians. Since the late 1950s Nairobi had been at the hub of the East African tourist industry. With its large airport and modern hotels, Nairobi was the starting point for the great majority of tourists on 'safari'. There, tourists spent their money on hotels, outfitting, and other tourist mementos, and supported the safari firms. Tanzania had the better national parks so the tours would swing south in a loop through Serengeti, Ngorongoro, and Lake Manyara before returning to Kenya. Tanzania's portion of the tourist dollar was confined to park entrance fees and a few nights' hotel bills, a mere 15 per cent. Consequently when the border was closed Tanzania saw its opportunity to declare that tourists could start their journeys in Arusha.

The lesson was learned sure enough, but not quite the one that was envisioned. There is nothing quite like roughing up the tourists to teach every Western travel agent, tour company, not to mention potential tourist that they are not welcome. Tourism stopped dead in its tracks, quite literally overnight. In 1976 there were 76,000 foreign visitors to Serengeti. In 1977 this number plummeted to 12,000 (nearly all were those living locally) and it stayed thus for nine long years.

* * *

The border closure presented us with a formidable logistics problem in carrying out the research on the Serengeti populations—in finding out what had happened to them—for now we could not fly to count them, nor even cross the border to reach them. We were going to have to find a new way to operate.

I had made arrangements to conduct the census with the Tanzanian National Parks and the Chief Park Warden, David Babu. They had given their enthusiastic consent. We were to use Mike's Cessna 182 aircraft, recently brought out from England, and still with its UK registration. That was our first asset because with such a registration the Tanzanian authorities may allow it across the border. However, that had still to be determined in the next four weeks. My job was to go ahead to Serengeti and make arrangements for the census. This covered everything from obtaining aircraft fuel (Avgas), laying fuel depots at remote strips, and finding and plotting the distribution of the wildebeest, to bringing in food and arranging our accommodation. Normally we would bring all our supplies in from Nairobi across the open border at the Serengeti–Mara boundary. Now, with the border closed, none of these supplies were available.

Two days later I flew to Dar es Salaam. It took five days to reach Seronera, first by bus to Arusha, and then to Manyara; and finally by Land Rover to Seronera reaching it at 5.30 in the morning.

* * *

Having finally reached Serengeti I could at last focus on science. I set out to find the wildebeest, demarcate the area they were in, count the calves from a sample of animals, and observe their behaviour. One of the most distinctive features of the Serengeti migratory wildebeest is that they produce their calves over a very short period of time. Within approximately one month some 90 per cent of calves are born, usually around February and March. Other large mammals such as caribou in the Arctic tundra have an even shorter birth period covering some ten days. They are able to do this through a very short mating period in October, the timing being set by the rapidly declining day length in those regions, to which the caribou are sensitive. But how do the wildebeest synchronize their matings when day length hardly changes through the year? The maximum difference in day length is only about twenty minutes.

Over many years in the 1960s and 1970s I had recorded when wildebeest gave birth and combined my records with those Murray Watson, who had studied wildebeest in the early 1960s. From these records one can calculate when calves were conceived since we knew the conception period. Plotting these conception dates year by year a picture emerged that gave a clue as to what was happening. Each year the conceptions started ten days earlier than the previous year, sometime in May. After about three years the conceptions then jumped to about a month later, and the process started over again. What could trigger conceptions ten days earlier each year? The moon. The full moon, for example, shifts ten days earlier in the month each year, a phenomenon called precession. Of course it may not be the full moon that acts as the trigger, but some phase of the moon seemed to do so. However, there are twelve moons in the year so how do the wildebeest know which one to respond to? It turns out that the May full moon occurs when day length is shortest, or perhaps more probably when night length is longest at the new moon. In 1977 I was putting this information together for publication[184] and needed to check the predictions on the timing of conceptions by observing the mating period, called the rut.

The wildebeest were obligingly all out on the plains, just right for aerial photography. After a month of observations on the rut, all was ready for the census. Or so I thought.

10

One Million Wildebeest

20 May 1977. I had been in Serengeti a month. Aviation fuel had been purchased in Mwanza and, with the help of David Babu, the Chief Park Warden, a truck had transported the ten drums to park headquarters. The rains had been good and the wildebeest were on the plains in a perfect distribution for our census. All that remained was for Mike Norton-Griffiths to show up with his plane on the day. We had arranged this as I left Nairobi because we could not communicate after that. I waited at the Research Institute, staying in a house next to the airstrip used by the Institute ecologist Jeremy Grimsdell.[185] It was easy to see a plane arrive—except that it didn't.

Next day the waiting began again. This was the part that we could not plan; nobody knew if the Tanzanians would let through a UK-registered plane—we had hoped that since the quarrel was with Kenya not Britain, they would not mind. As dusk approached I was becoming convinced that the authorities had not cooperated, the trip was for nothing. Then a distant hum, becoming louder, got my hopes up. Sure enough it was a plane. It had to be him because all others had been grounded. He roared over the house, banked, landed, and taxied to within 100 yards of us. I could not have been more delighted not just because of the census, but because he was also my only way of getting out of Tanzania and back to Nairobi.

Mike had arrived at Kilimanjaro airport the day before, as planned, to pass through immigration before flying on to Seronera. Although Kilimanjaro was an international airport, his was the first plane to land there in the whole month since the border was closed. The runways are designed for Boeing 747s and miles long. The control tower gave him clearance, and presumably because the whole airport staff, including immigration officers, were bored out of their minds, they all came out to watch him land. Not thinking, he landed at the beginning of the runway, as we always do, and then realized he had miles to taxi. So he took off again and landed further down the runway. For some reason this did not impress the control tower.

After he had parked, the immigration authorities seemed unimpressed too: by bad luck one of them had been an immigration officer at Seronera some years earlier and there had been some form of conflict. He decided to get his revenge. Mike's explanation that he had come to count wildebeest seemed far-fetched. The officer implied that Mike was something of a cowboy pilot. So Mike had to stay at the airport for the night while they checked up on him. He had, after all, come from Kenya and could be a spy in disguise. Luckily, by the next morning the authorities seemed to be satisfied with Mike's story and let him go.

* * *

Mike and I began the process of setting up the plane to take vertical aerial photos of the wildebeest. This entailed fixing a motorized camera to a wing strut facing directly at the ground. It was operated by a chord that came into the window of the co-pilot, which in this case was me. Then we took off and flew around the Serengeti plains marking on a map the edge of the wildebeest distribution; at this season (the end of the rains in May) they were in one huge herd. Back home we then drew on the map parallel transect lines through this distribution—these were to be our flight lines next day. The lines were effectively west–east and crossed the main road from the park entrance to Seronera.

Meanwhile, unbeknown to us, international affairs were proceeding rapidly. Rumours of all sorts abounded in the newspapers, including gun-running across the border in Serengeti and troop build-ups on the Kenya side of the border. Most likely untrue, these rumours were nevertheless taken seriously by the Tanzanian Government who ordered troops to the northern Serengeti.

The next day dawned calm and clear, just perfect for aerial photography. We took off, climbed to 1,000 feet, and set out to fly back and forth systematically moving south and taking our photos. At one point I noticed far below lines of trucks moving slowly north. I commented on them to Mike, pondering what the trucks were doing. We had seen nothing like them before. We kept crossing over them while we were taking photos of the wildebeest as the line of trucks crawled along.

The census was an excellent one. We were well pleased as we came in to land at the research airstrip not far from our house. Strangely there was a reception committee to meet us. There was a high-ranking army officer, several troops with menacing-looking guns, and a very nervous park warden. The officer approached us and demanded an explanation for what we were doing as we were flying over them. Cheerfully, at first, we replied that we were counting wildebeest. He clearly did not believe us, asking how we could count wildebeest when we were flying so high. I replied that we took photos and counted them later. The officer's tone became distinctly chilly, pointing out that we had been flying over his troops taking photos repeatedly. Suddenly all became crystal clear. For whatever reason they were there, we had been systematically photographing the convoy. With an expanding sense of desperation we appealed to the warden, who confirmed we were indeed photographing wildebeest. The officer turned to Mike and asked where he had come from. Mike rather too hastily replied that he had flown in from Kenya. 'So', concluded the officer, 'you are from Kenya and you are photographing our army.' It was clear the wildebeest story was a rather pathetic ruse. 'I am arresting you for spying for Kenya.' We were to remain under house arrest while he impounded our plane and camera until he decided what to do with us.

We retired to our house not far away and an army guard was stationed at the plane. Nobody touched the camera (which was attached to the wing strut) and I managed to secrete the exposed films into my bag as I left the plane to take to the house. There we stayed for two nights awaiting the outcome; Mike was despondent at the prospect of losing his plane. On the third day we began to notice a pattern of events at the plane, which was visible to us from the verandah of our house. Every three hours or so there was a change of guard at the plane. The army base was at Seronera 3 miles away. The guards had to walk from the base. However, instead of the new guard coming from the base and sending the old one back, the reverse occurred. So there was a gap of an hour or more when there was no guard on the plane. Once we had appreciated this, it took but a moment to realize our next move. We packed our bags and waited for the next guard change, which was about 5 p.m. As soon as the guard had gone sufficiently far down the road not to be able to run back, we hastily carried our bags over, stowed them, and took off. We were jubilant, we had escaped.

Well, not quite. By the end of the census we had used up most of our fuel. We were short of avgas, it was late in the day, we could not go far, and we could hardly come back to refuel. Our only hope was to fly the thirty minutes to Olduvai, stay the night at Mary Leakey's camp,[186] and hope that she had some fuel.

Mary Leakey was most welcoming, and yes she had some avgas, just enough. She was, however, not well and she promptly asked us for a lift to Nairobi. No problem, Mike said, until it dawned on us what that meant—we would have to go out via Kilimanjaro airport so that Mary could exit officially. We had not planned on doing this, of course. We stayed the night at Olduvai camp with various archaeologists and Mary proudly showed us the first casts she had made of footprints at a nearby site called Laetoli. 'Do you think these are human?' she asked us. To our untrained eye we thought they looked completely modern. They were probably 3 million years old, she announced, but they were doing more exact dating at that moment. Those casts were breathtaking.[187] Strangely juxtaposed amongst all these people was a semi-tame cheetah that wandered in and out of camp at will.

Next day the three of us took off for Kilimanjaro airport an hour away by plane. Our arrival was uneventful and Mike and I cleared the immigration and customs with no problems, although they remembered Mike from his arrival a few days earlier. Then it was Mary's turn. As soon as they saw her they took her into a back room while we waited anxiously, visions of impending doom descending from the military. After an age she reappeared smiling. She asked Mike what kind of flying he had been doing recently. The authorities, she explained, thought Mike was not a person she should be flying with and they were trying to dissuade her from going with him; they were concerned about her safety.

With considerable relief we raced for our plane, and watched anxiously as it was refuelled, which seemed to be proceeding in slow motion. At last we were in the air, and as we crossed the border we cheered: we had finally escaped.

* * *

The count of the wildebeest from the photographs a few weeks later gave us a considerable surprise. The population was 1,400,000. This was double what we had counted four years earlier. It was also now the largest ungulate population in the world. Clearly the ecosystem was still changing rapidly.

This result provided the answer to several important questions concerning the ecology of Serengeti. First, it explained the changes that other scientists had noticed in nearly every aspect of the system—the frequency of fires in the dry season, the survival of trees, the population trends of other ungulates, and changes in the lion population. These changes had been documented by different scientists but their causes had remained speculative.

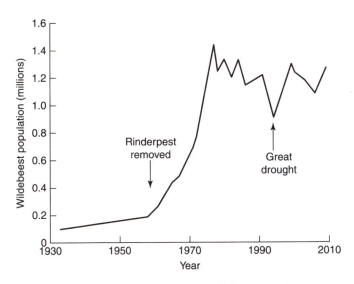

Figure 1 Changes in the Serengeti migratory wildebeest population since 1933.

It was now clear that the wildebeest were the keystone to the whole ecosystem. This was the answer that made sense of the whole ecosystem and would be the main thread of our synthesis.

Secondly, the trend in the wildebeest population was the test of the facilitation theory of migration. If wildebeest were providing a niche for the Thomson's gazelle then an increasing wildebeest number should allow an increasing gazelle number. In contrast, if interspecific competition was acting, then gazelle numbers should decline. The count of wildebeest was such a surprise that we repeated it the following year (1978) with much the same result. We also counted the Thomson's gazelle. Their numbers had almost halved since the previous count in 1971 when the wildebeest were less than half what they now were.[188] The answer was clear: competition was a stronger process than facilitation. The earlier results[189] could be explained as a transient effect during the period when the migration was moving off the plains. But this was before the dry season when competition took over.

Thirdly, the increasing wildebeest numbers meant that regulatory processes slowing down the population, which had been seen in the buffalo population (Chapter 8), had not yet been seen in 1977 in the wildebeest.

As a fitting postscript to this story, Mary Leakey went on to publish her ground breaking discovery of the footprints of early humans at Laetoli a year later, footprints made some 3.2 million years ago.[190]

11

Outbreak of Trees

THE trees were disappearing in Serengeti. This was a phenomenon apparent all over savanna Africa in the 1970s and the general conclusion of
scientists in Uganda, Kenya, and elsewhere was that elephants were the cause
of the decline.[191] However, an alternative theory was developing in Serengeti.
In the early 1970s Mike Norton-Griffiths and Harvey Croze[192] had been
studying the causes of the tree demise and had concluded something very
different: the high frequency of fires was preventing the regeneration of trees
and so over several decades the slow death of adult trees was not balanced by
new young trees entering the population. Photographs taken in 1944 by the
professional hunter Syd Downey of the western side of the Mara Reserve in
Kenya showed extensive woodlands of acacia trees. This was part of the
northern Serengeti ecosystem. When we repeated these photographs in 1984,
forty years later, only two trees remained in the same area.[193] Aerial photos
of the Serengeti taken in 1959, 1966, and 1971 showed a consistent decrease in
the tree population.

By 1980 I realized that the two theories for the disappearance of trees
could be tested to sort out which was correct. If elephants were the cause
then we should see a continued decline in adult tree numbers since elephants were still present in large numbers. However, we knew from the
studies of Mike Norton-Griffiths that the extent of burning in Serengeti
had dropped dramatically over the past twenty years and we had realized
that this drop was due to the increase in wildebeest numbers. They were eating
more grass, which was therefore not available for burning and so fires were

not spreading as far.[194] We predicted that if fires, leading to a lack of young trees, were the cause of the decline then we should be seeing a pulse of new trees appearing with the drop in burning and, therefore, an increase in the tree population.

The outcome was clearly important for the conservation of Serengeti. If Serengeti was going to change from savanna to open grassland there would be changes in the large herbivore populations—not only would elephants run out of food, but giraffe, eland, and even impala would also be affected. Birds would be even more disturbed because a large number of species depend on trees for insect food, for perching to see insects in the grass, and for nesting. Trees also alter the soil properties for holding water and the flow of important nutrients such as nitrogen, phosphorus, and calcium that are required by the mammals in their food.[195] The disappearance of trees would clearly alter the whole ecosystem, maybe the great migration, and so in the end the conservation status of Serengeti itself.

To test the theories I set up a photographic monitoring programme in May 1980.[196] I took photographs from the tops of hills of the woodlands below. In those times when flying was banned, fuel was hard to find, and the technical ability to interpret aerial photos was becoming lost as educated Tanzanians left for other countries, it was clear that I needed to use a technique that was simple, easy, and inexpensive—one that anyone could repeat and required no advanced training. I hoped my photos could be repeated in fifty years' time using ordinary hand-held cameras.

The problem with photos taken from hills is that they provide oblique views, which means not only that the area they cover is a distorted polygon—a greater area in the distance than in the foreground—but that it is virtually impossible to measure the area on the photograph. Hence oblique photos are not usually used for photo-analysis of vegetation. However, I reasoned, all we were interested in for this present purpose was to see the relative change in the number of trees. If we used exactly the same area on photographs taken from the same place at different times we did not need to know the area, only the number of trees in that same area. For example, if there were 50 trees in a photo in 1980 and only 25 in the same area in 1990 then we can say there was a 50 per cent decline over ten years. And that was what we were interested in.

So obliques would do just fine since we could draw a line around obvious markers that appear in each photo to set the same area.

By good fortune Serengeti has a number of steep-sided ironstone hills, some at 1,000 feet above the generally flat or gently undulating savanna (see Chapter 2). They are stony, some have rocky bluffs, and many have sharp, knife-edged tops, just the features from which to take photos of the plains below. I set out to walk up as many of these as I had time for, marking the point by using a cairn of large stones with a metal pole placed in the middle, this being well before the days of GPS recording. On that initial trip in May–June 1980 I managed thirteen hills or kopjes. Photographing the landscape from viewpoints on hills may sound straightforward, but nothing is that simple in the bush.

* * *

Luca, as I shall call him, was a gentle and genial soul, only too keen to show me the traditions of Tanzanians. He took me around the bazaars in the towns, the markets in the villages, showed me the different porridges they made from different millets, and introduced me to many of his friends. He was a bona fide good chap.

Luca was trained as a biologist at the University of Dar es Salaam, his passion was horticulture, and he wanted desperately to go into the Department of Agriculture. Instead, he was assigned by the government regardless of his wishes, as was the custom in those days, to the Department of Wildlife, from which he was sent off to Zanzibar to study primate behaviour for his Master's thesis. Having accomplished this successfully, he was again disappointed, indeed horrified, at being appointed Park Ecologist in the Serengeti National Park in 1980. He knew little of ecology, even less of Serengeti. His great phobia was lions; the very thought of them made him tremble.

Luca was assigned to me by the Chief Park Warden, David Babu, to be trained in ecology, wildlife management, and the needs of the Serengeti. He was to accompany me on my walks up the hills; a good way, explained the CPW, to get a feel for the environment, the vegetation, and the animals. Luca was not convinced, considering it a good way to put oneself in harm's way.

We were to climb Banagi Hill, some 700 feet up. As well as Luca there was Monique Borgerhoff Mulder, an anthropologist who had recently arrived[197] and who had asked to come with us. This was an opportunity to get out of the car, see something on foot, and get exercise. We climbed the shoulder of the hill, it being less steep and clear of thickets that hid buffalo, who liked that hill—we knew that well enough from our days when we had lived at Banagi ten years earlier (see Chapter 3). Thin stands of whistling thorn clung to the rocky ground;[198] aloes were common and, when flowering, added colour to the grey landscape. We left early, 7 a.m., to avoid the heat and haze so as to get clear photos.[199]

We had left the vehicle at the bottom, below the steep side. We could see it far below, looking like a little children's toy. After completing the photography we set out to walk down the steep side. I was ahead, the other two some 50 yards behind, talking. A sudden deep growl and snarl preceded a row of yellow manes and teeth. Lions rose immediately in front, just 5 yards ahead, threatening, snarling, teeth bared. I leapt backwards, it was a big leap, yelling 'Simba' (lion). I backed away, then turned and saw Monique running. I caught her up, enquiring where Luca was. He had gone, we saw him far ahead of us. The lions, having got rid of us, ran straight down the hill. We traversed along the hill for a while and finally caught up with Luca. We had a clear view of the lions, still running far below. There were six of them, two males, four females. Luca exclaimed, while looking anxiously down the hill, that there was a whole herd of them and that they were waiting to eat him. The lions, having put distance between us, had slowed down and made for the only substantial shade available, which was our vehicle, the whistling thorn not passing as shadeworthy. They settled down in the shade next to the car.

With some exhortation and cajoling we persuaded Luca to walk down the hill with us towards the vehicle—and the lions (which were not really a threat). And as we went he dragged further behind, his whole being telling him he was walking into the jaws of death. As we approached the Land Rover the lions, who of course had been watching us, got up and ran to the side and back up the hill to avoid us, two to one side, four to the other, and made to rejoin each other, neatly forming a pincer movement—and coinciding with

Luca's position a 100 yards or so behind us. Poor Luca, his worst fears had materialized: they were coming to get him. If Luca tarried before, he made up for it in that last 100-yard dash. Impressive it was. Never mind that, the lions were actually heading in the other direction.

* * *

A few weeks later we were in the far north, Luca and I, next to the Kenya border. Bandits were roaming the country, fifty at a time. They had AK-47 semi-automatic machine guns. The ranger post at Kogatende had a siege mentality. There was no vehicle, the rangers had to patrol on foot, but because they were vastly outnumbered and outgunned they were loath to do even that—they stayed in their post and rarely ventured out. The lack of vehicles also meant that the tracks had become overgrown with the 6-foot high *Hyparrhenia* grass common in that country.

Our objective was some kopjes about 5 miles away.[200] This is the most attractive part of the whole Serengeti but it was also the most difficult to traverse. We had to cross several deep ravines, and open rolling plains with scattered trees. The grass was long and there was nothing for it but to walk. Luca was not happy because he could see very little of what was in front of him in that grass. He asked if we could take a ranger with us but I had failed to persuade any of them to come with us. Luca, when motivated, was hard to ignore—and after some serious debate with the rangers, one was assigned to come with us. I believe he was the least popular of the squad, or had otherwise earned this as a punishment. Anyway he was less than enthusiastic when we set out next day. He started by saying we should not go, explaining the dangers of the bandits in lurid, but almost certainly true, detail. This merely terrified Luca more, and he made certain that the ranger was not going to escape—Luca had come to know me better: there was no going back, so the ranger was coming with us like it or not.

We set off, the ranger in front. He carried his Lee Enfield .303 First World War rifle over his shoulder with a round up the spout, waving it around and mostly pointing it at me just behind him. I felt a lot more threatened by him than by anything else. We stopped shortly after starting. I insisted he give me his six rounds so that he was not tempted to use one, especially on me. He objected and so did Luca. This debate raged for awhile. A compromise was found whereby he kept the rounds in his pockets.

We set off again, the grass flicking across our faces, small animals starting and running unseen ahead of us, almost certainly oribi, which were abundant,[201] birds fluttering up from nests in the grass—white-browed coucals, rosy-breasted long-claws, red-shouldered widowbirds, Jackson's whydahs, cardinal queleas. Sometimes gamebirds would rise startled, cackling and flapping, making us all jump—red-throated spurfowl and helmeted guinea-fowl they were. Numerous insects were crawling over us, picked up from the grass brushing across us: stick insects, caterpillars, bugs, grasshoppers, ants of many species. They climbed into our clothes making us scratch incessantly. Ahead of us we saw a lone bull elephant peacefully grazing. It was right in our path but upwind so it had not noticed us. We circled around it, giving it a wide berth, before continuing on our way.

Closing on the rocks that we were aiming for we paused to look with the binoculars. All was well. We approached the rocks, circling to find the best way up, and it was then that we saw the lions, females and cubs sitting up now, watching us below. The ranger was unperturbed, merely clapping his hands to frighten them off—and they duly obliged us by running down the other side, into the long grass where we could no longer see them. For the second time we looked around and found Luca had vanished. We then had to walk around calling for him until we found him up a tree on the far side—the same side that the lions had descended. Yes, he said, he had almost bumped into them, and had shot up a tree. Fate, it seems, had decreed that Luca and lions were to meet. Luckily, from his perch, he could see that they had headed for the river.

Having taken the photographs from the top of the kopje we set out to return. The same elephant was still there, but now downwind and he caught wind of us from a distance. These were very frightened elephants; they had experienced years of ivory hunting by the bandits (see Chapter 15) and so were very aggressive. He turned, faced us, and then started to run towards us, ears out, trunk up testing the wind. We were in almost open grassland (the reason we were taking photos of course) and no place to be with a terrified elephant. This time there was nothing for it but to run, to one side where the river was half a mile away, and where there were banks and trees to hide in. We made it, breathless, sweating profusely in the midday sun. The elephant had given up early but we were taking no chances. So we gingerly made our way down the riverine forest until we judged we had passed the danger before resuming the walk back to our vehicle an hour or so away. Luca did not enjoy these outings much; they were somewhat of an ordeal.

* * *

Of all the many unexpected events that took place in our studies of the Serengeti ecosystem perhaps the greatest surprise was what happened to the trees. Their decline over some fifty years had of course been the motivation for setting up the photopoints. According to the prevailing theory that elephants were destroying the trees there should be none left in a decade or so.

In fact the evidence that trees were regenerating was already becoming apparent but I was not looking for it because I was expecting to see the disappearance of trees. I repeated the 1980 photographs in 1986 and there, undeniably, were thickets of young trees. By 1991 the evidence was clear to see—there was an explosion of small trees everywhere I had set up photopoints in Serengeti. This applied to all the common species, both African acacia species and the broad-leaved *Terminalia* and *Combretum* species of the north-west, our version of the miombo woodlands of southern Africa. By the year 2000 these had developed into a dense stand of trees, so dense that in many areas we were no longer able to drive through them. Some of the smaller species such as the whistling thorn had already grown into dense stands by 1986 (see Chapter 14), catching us by surprise.

The results were confirming Mike Norton-Griffiths' theory that fire was the determining factor for tree populations in Serengeti. When the frequency of burning declined in the 1970s due to the increasing numbers of wildebeest removing the grass (so that fires had less fuel to burn) there was less damage to young saplings and they started to grow up. Elephants, it seemed, were not the culprits that people had thought they were.[202]

12

Sudan

SERENGETI, initially known for its lions in the 1920s, became famous for the great migration of wildebeest and zebra in the 1960s. Since then scientists have come to understand how the wildebeest migration determines the whole ecosystem. What causes a migration? What are the consequences of migration for the animals? Since very few species migrate, especially amongst mammals, why do some migrate and most others not? Clearly we must find answers to these questions if we are to understand why Serengeti is unique.

In 1962 Murray Watson was the first person to record accurately the areas that the migrants used at different seasons. But the records were piecemeal, showing location but not accurate numbers. So Mike Norton-Griffiths and his associate, Linda Maddock, decided in September 1969 to set up a regular aerial survey of the ecosystem and document more precisely what was happening. This was a mammoth task. Each month a plane flew over the whole area, some 10,000 square miles, in straight lines 7 miles apart. It took two days of flying twelve hours a day. Two teams of four alternated in six-hour shifts. We did this for nearly three years. By the end of it we had a better documentation of the whole system than either before or since, an enormous wealth of data that we are still using today. It provided the example for the scientific monitoring of ecosystems.

From these data it became clear that the migrants were following the rain.[203] In the wet season the migrants were on the far eastern plains where, perversely, the rain was least but the grazing was best for animals that liked short grass. However, these plains dried out first and so the animals had to leave and they moved towards higher rainfall areas. As I have described in Chapter 2 there is a gradient in rainfall with the least falling on the south-east plains and the most occurring in the north-west woodlands. So, as the dry season sets in, the animals simply move up this gradient.

However, this discovery raised the question: why don't the animals stay in the wettest areas all year round, like some of the other non-migratory species such a buffalo, kongoni, and impala? The answer to this seemed to be food. The first plant ecologist working on the Serengeti, Hubert Braun from the Netherlands, had shown in the 1960s that grass on the plains was high in nutrition.[204] From this observation I had realized that migration might be an adaptation to make use of temporary good food. If animals can move, they can go to the best food in the ecosystem, fatten up, and then move back again when conditions become too difficult to stay there. Only those species that can eat that type of food would benefit from migration.[205] This theory of migration had to be tested and to do so we needed to study another system.

* * *

In the late 1970s another great migration of mammals was discovered in southern Sudan. Civil war in Sudan had been raging through the 1960s but a peace treaty of sorts had been agreed in 1972 between the Arab north and the African south. The calm that followed allowed conservationists to survey the country. This migration involved a beautiful antelope, the white-eared kob, which likes to live in wet plains and swamps. The Wildlife Conservation Society of New York was interested in finding out more about this mysterious species—how many were there, where did they go, how safe were they? We did know their numbers were huge and that they travelled a long way. It seemed that this was a migration similar to that in Serengeti, an astounding find if this were true. It was also an opportunity to see whether our understanding of how migrations worked, derived from the Serengeti research, still held true in this one.

Sudan was no place for the faint-hearted. Remote, out of touch with the modern world, cut off from supplies, the south-east corner of Sudan, called

the Boma, is a vast area of flat grassland, scorching hot, an impassable quag-
mire in the rains, and bone-jarringly rough in the dry season. The soil was of
silt, locally called 'black cotton', which baked hard as rock in the dry season
and broke into cracks wide enough to take half a wheel—progress was a
painful 5 miles per hour that destroyed vehicles and caused them to overheat.
At the first heavy rain this silt turned immediately into a bottomless glue and
vehicle progress slowed to zero. Walking also became impossible as the
gumbo accumulated on boots with every step so that after about ten the
weight was too heavy to raise one's foot. It was little wonder that wildlife still
survived—no one could get in to destroy it.

The puzzle was how the wildlife managed in such surroundings—and
there was much wildlife, for besides the white-eared kob there were also good
numbers of hartebeest, topi, zebra, Thomson's gazelle, giraffe, and elephant,
and even a few herds of buffalo. In this remote area there lives a hunting tribe
called the Murili. The men were usually naked and hunted kob with dogs and
spears; they lived simply in temporary grass huts, moving with the kob as
they migrated. The women decorated themselves with lip-disks—large
wooden disks the size of an ashtray inserted side-on in the lower lip so that
they protruded several inches out from the face. A hole was made in the lower
lip and then over time gradually stretched across ever-increasing pieces of
wood until eventually the lower lip protruded outwards. This was one of the
last tribes to continue this practice. It was in many ways still the Africa of the
eighteenth century.

George Schaller phoned me one day in 1979 and suggested we find a student to study the migration of white-eared kob. John Fryxell, whom I had
known as an undergraduate, was visiting from California, and immediately
expressed an interest. A few months later John departed for Nairobi, and in
due course went to Sudan. The main supply town was Juba, which is on the
Nile. From there he drove for two days over barely visible tracks to the headquarters of the new Boma National Park. This park was being developed by
the Frankfurt Zoological Society, and John's job was to document the ecology of Boma, an area of some 10,000 square miles.

* * *

In February 1982 I visited John to see how he was progressing. I arrived from Nairobi in a small plane on a late Saturday morning at Juba, a dilapidated and smelly town on the banks of the Nile. It had once been a thriving town but had decayed from years of neglect. The temperature was well over 100°F. There was no one at the airport to meet me, and no transport either, although the strip was several miles out of town. Eventually I hitched into the town in the back of a truck, hardly a good way to start. There were almost no places to stay and it was several hours before I found a hostel with a bed, a broken toilet, and no food.

On the Monday, after talking with the officials in the Wildlife Department, I met up with the recently appointed Park Warden of Boma, who by chance had flown to Juba for supplies. He flew me back to Boma headquarters some 200 miles east on the Ethiopian border. This was one of the most remote places I have been to in Africa. The headquarters consisted of three tents, several grass huts for the local labourers, and some German army rocket launchers. One needed a crane to change a wheel and as for mending a puncture it was not to be contemplated—one hoped it would never happen.

It took a week to find John; we flew out each day to search for him. Eventually we tracked him down by following in the plane some tyre tracks in the grass for 20 miles. We saw his tent and I dropped him a note to make a rendezvous next day at a patch of open ground where we could land. It was the end of the dry season and we were in the north-east, a remote and lonely place. Supplies were limited and Juba was a week's journey away. We shot our own meat but water, vegetables, and indeed everything else had to be measured out very carefully. We soon learnt to bake bread in a tin can covered with embers, extract honey from wild African bees' nests, and take advantage of opportunities—one exotic meal was made of a wild gargany duck killed by a Lanner falcon in midair. The duck dropped at our feet out of the heavens, and although we moved away to let the falcon claim its prize, it was too timid and flew off. So we gratefully accepted the present. For the most part we ate fillet of white-eared kob.

From the air I had seen the remote swamps in the far north of the system with vast numbers of kob. Until then, no one had known where they went during the worst of the dry season, which we were then experiencing. We needed to see these swamps from the ground, and so we made our way there across country some 30 miles. We had a Toyota landcruiser and a German Unimog four-wheel-drive truck. The ground was dead flat, covered with trees and bushes spaced about 50 yards apart. We could not see far and everywhere looked the same—it was easy to get lost and we had to go by compass most of the time. The swamps were wonderfully green after the dry brown and burnt country we had been in, and there indeed were the kob, hundreds of thousands of them, and everything that eats them—lions, hyenas, leopards, and Murili hunters.

A few days later John had to return to his base camp to get more supplies. That night we had a severe rainstorm and the ground turned immediately into a bog. I could not move, but worse still John had not got back to his base camp—I knew because I had a radio transmitter and talked to his African assistant there. Two days later John finally crawled into his base; he told me he could not make it back again in his Unimog. I had to complete the work alone and drive out when the ground firmed up. I spun out one and a half days of food into four and boiled up stagnant swamp water using grass as fuel, for there was no wood in the swamps. It was lonely but beautiful—thousands of water birds of every sort: herons, storks, egrets, ibis, ducks, geese, waders, pelicans, and one of the rarest and most peculiar birds in the world, the shoe-bill stork. It lives on frogs and is found largely in the Sudanese swamps, with a few in Zambia and Tanzania.

* * *

The Frankfurt Zoological Society (FZS) was funding the development of Boma National Park for the Sudanese. The director of FZS, Dr Faust, decided to bring a party of wealthy patrons on a safari to the Boma swamps. They commandeered a rocket launcher from the German army, installed seats on top, and lockers and beds below, shipped it to Mombasa, and proceeded across Kenya, then southern Sudan. John and I were at our base camp when we received a message by radio that they were coming. We arranged to meet in the middle of the Boma woodlands and lead the party to the kob migration in the swamps. This monster machine had countless wheels and so negotiated the rough ground with ease, but it was rather heavy and as we approached the swamps it began to sink into the ground. Faced with the prospect of never getting it out of the mud, we decided to stop for the night and proceed in our small Toyota next day.

The Murili hunters had never in their lives seen such a machine. They were naturally very curious, in fact just as curious as the good German patrons, many of whom were women, for in short order the truck was surrounded by naked hunters. We witnessed the unusual sight of fully dressed tourists trying to take photographs but appearing rather embarrassed, as each one was surrounded by inquisitive and completely unabashed hunters. It was quite clear who were the watchers and who the watched. As the visitors unloaded their huge assortment of stuff for the night, the Murili simply sat down and watched them, following them off to the bushes when one of them needed to go to the toilet or strip to wash, and causing all sorts of etiquette problems. I wondered whether any of them realized that this is what it is like when hordes of tourists in minibuses descend on tribal villages. We cooked our meal and sat around the campfire telling stories with the tourists, watched all the while by the Murili. As we retired under our blankets on the ground so the Murili curled up and went to sleep with us.

The morning began at dawn when the tour group was awakened by the howl of the Murili dogs encircling a kob in the middle of camp and the hunters spearing it to death, blood spattering everywhere. They then proceeded to dismember the carcase and eat it while the tourists were trying to negotiate their way amongst them, wash, clean their teeth, enjoy their breakfast, and generally pretend that it was like a normal day in Frankfurt.

Dr Faust and I left these entertainments and headed for the swamps. There we saw one of the great sights of the natural world: antelopes as far as the eye could see. They were feeding on the edge and in the swamps as well. There were countless water birds, and in amongst them we saw the shoebill stork, which is what Dr Faust had come for. It was a memorable occasion as we walked into the marshes to confirm our sighting, shaking hands over our delight. It made the trip for him.

John and I accompanied Dr Faust and his group out of the swamp and set them in the right direction for the Boma headquarters a long way south. Back at our base camp we resumed work. This camp was on one of the paths used by the Murili in their travels, and they often stopped to watch us. We were clearly a novelty and being short of entertainment they would sit and watch, sometimes for a day or two. When we cooked up our kob stew they joined in. We learnt to make enough stew for unexpected passers-by. As usual they spent the night curled up where they sat. One night we had a rainstorm; it was very heavy. In no time I found my tent invaded by about ten naked Murili, all talking excitedly. They sat down next to my bed and happily went to sleep again. In the morning they left as if this was a perfectly normal night's rest.

* * *

August 1983. Sudan erupted into civil war again; the African south led by General Garang was in rebellion against the Arab government. John had finished his fieldwork that month and was on his way back to Juba. He was camped some miles south of Boma when a messenger arrived to tell him that Boma headquarters had been captured by the rebels and that he must not go back. His replacement as ecologist for the project, Conrad Aveling, had just arrived and was promptly captured. John continued on his way to Juba; fortune was smiling on him.

Boma headquarters was situated at the base of a steep escarpment, the edge of the Boma plateau, which was part of the Ethiopian massif. I had made a trip up there into highland forest, cool and green in comparison with the plains below. There we found wild bushes of coffee for this was the area where coffee evolved. The patches of forest contained birds not found elsewhere but the patches were being cut down and would soon be lost. I wanted to record what was there before they disappeared. At the top there was a mission station and medical dispensary run by an American missionary. The mission was also captured, the rebel army taking the Boma personnel up the escarpment.

Meanwhile in Juba things were slowly grinding into action. The official (northern) Sudanese army was planning to attack the rebels and rescue the mission—John was consulted on the geography and the logistics of this operation. The army was able to communicate with the mission because the doctor was still conducting his medical work and had the freedom to talk by radio. He was warned that on a certain night the army would climb the escarpment and attack the rebels; he, his staff, the Boma staff, and the civilians attending the clinic should hide.

Many if not most of the rebels had ailments and they were lining up for treatment at the clinic each day. The doctor organized a clinic for the set night and handed out his customary tablets to the rebels—except that on this occasion they were barbiturates. In due course all of the rebels fell asleep, the prisoners walked out of the mission and hid in the forest, and waited. They waited all night and nothing happened—no Sudanese army in sight. By morning the rebels had all awoken, and seeing their prisoners had fled set out to search for them. It was during this period when rebels were chasing prisoners around the forest that the army finally appeared and promptly started firing randomly. Many of the rebels were wounded as were some of the captives— now trying to take cover from the mayhem around them and in serious danger of being killed by their rescuers. Eventually the army won, the mission was freed, but most of the rebels escaped.

* * *

The million dollar investment by FZS in constructing the Boma Park was almost entirely lost. I had discussed with Dr Faust the previous year, when we were together in the Boma swamps, the possibility of FZS returning to Serengeti to rescue it from the collapse in management. He had explained that they were committed to Boma and they could not run two places. After the destruction of Boma Dr Faust contacted me and said they would return to Serengeti. It was the best news I had received for many years. In due course Markus Borner, the FZS representative based on Rubondo Island in Lake Victoria, moved to Seronera and began to repair the institutions and infrastructure, a job that was to take another fifteen years.

As a postscript to this story, the second civil war finally reached a truce after almost a quarter of a century in 2005. General Garang had led the south for the whole period, but mysteriously died in a plane accident almost as soon as the truce was arranged. A vote for separation was conducted in January 2011 and South Sudan officially became an independent country in June 2011. Scientists flew over the Boma in 2007 and reported that the herds of kob were still there—as we expected, the Boma is a natural sanctuary for wildlife, despite the caprice of human politics.

* * *

John Fryxell used his research in Boma to develop the general ideas about migrations, the reasons why they occur, and how they work.[206] In essence migration occurs when there are temporary sources of good food that animals can move to but then must move away again when those sources disappear. The extra food that animals obtain from this type of behaviour allows for large populations. We came to realize that all migratory populations of large mammals, such as caribou in Canada, bison in America, numerous examples in Africa, and even whales in the oceans, have large populations because they have access to this additional high quality but temporary food.[207] Migratory birds are probably also taking advantage of these temporary food supplies when they migrate to the high Arctic for the northern summer.

I had realized that in Serengeti none of the large predators could follow the migrant herbivores the whole way through the year—lions hardly moved out of their territories while hyenas moved only 30 miles or so. These hyenas lived on the edge of the plains and moved either further onto the plains or west towards the corridor. They went no further than that.[208] Did this mean that migrants escaped from their predators?

John Fryxell, using both the Boma and the Serengeti data, constructed mathematical models and made an interesting discovery. If the migratory prey numbers were very low then the predator population was able to regulate the prey population, in other words keep numbers at some low level. If, however, prey numbers somehow increased, then predators could not keep numbers down and the prey population climbed to a much higher level set by the food supply. The prey population could stabilize—at either a low predator-limited level or a high food-limited one. The system had two possible stable states.[209]

John Fryxell's discovery has important conservation consequences for it predicts that if a migratory population is allowed to get too low then it could get trapped at a low level by predators and may never climb back out again.

Have we ever seen such an event? Well maybe. There was a migratory caribou herd in the Yukon of Canada, called the 'forty-mile herd'. It was in the hundreds of thousands before the Second World War. During the war it was heavily persecuted, animals being shot by army personnel for food, and numbers declined to only about 10,000, possibly less than 5 per cent of their original number. Wolves are now the main cause of death in this herd, and although there has been much effort to restore the population it has never exceeded 14,000 animals. It appears now to be limited by predators.[210] It is possible that the same situation applies to the saiga antelope on the Asian steppes.[211] They were once a million or so in number but have declined to very low numbers from overhunting in the past twenty years. We will see whether they can climb back up again.

Our understanding of the mechanics of migration has another important consequence for conservation. Migrant populations can have large numbers because of the extra food they can reach by moving. This means that if they are ever prevented from moving they are automatically overstocked. They will eat out their now restricted food supply and the population will collapse. This has been observed wherever migrant populations have been blocked by fences as we have seen in Botswana and elsewhere.[212]

The fundamental conclusions from the research on migration are twofold. If the Serengeti is to survive, then, first, migrant populations must be allowed to stay in high numbers. And secondly, the migrants must not be blocked by fences along the edge of the park or along roads.

13

Coup d'état

B Y the 1980s scientists had documented the ecology of the large herbivores in savanna Africa and shown convincingly that each had its own
particular niche. This pattern conformed to the predictions of niche partitioning theory generated by interspecific competition. We had concluded in
our first synthesis of the Serengeti research[213] that competition was driving
events in that ecosystem.

There are, however, some problems with the concept that species with different niches can coexist (niche partitioning). If one looks hard enough at the
ecology of different species one is bound to find a difference. This is because
we have already identified the species as being different in the first place in
order to give them a different name—different in size, colour, behaviour, for
example—and so they will almost certainly have different ecology. In finding
different ecology (different niche) we have merely established what we had
already identified. This is a tautology, a circular argument.

So, the question becomes: how different do species have to be to show that
competition is allowing coexistence? Early ecologists had proposed that species had to be different by a ratio of at least 1:3—that is, no more than 30 per
cent overlap—if they are to coexist.[214] Since then that ratio has been shown to
be false, but none the less it was nowhere close to what we had been recording.
In essence, although scientists had shown differences between species in
their feeding and habitat choices there was a very large degree of overlap,
often as much as 80 per cent. Wildebeest and zebra, for example, ate almost the
same food, they lived in the same habitats, and they migrated together in the dry

season. Topi and kongoni are close relatives who ate much the same type of grass and lived in mixed herds. These patterns did not conform to competition theory.[215]

It was these anomalies that caused me to ask whether another process was allowing coexistence. Scientists had hotly debated in the early 1980s whether competition really was the driving force in shaping communities, or whether predation could explain the same observations.[216] In 1977 Bob Holt (who decades later joined us in the Serengeti work) had published a paper showing how predators could keep prey populations at low enough levels that they did not compete with each other. He called this theory 'Apparent Competition', which later became known as 'Predator Mediated Coexistence'.[217] Obviously, such species could overlap in their ecology a great deal. Was this process occurring in Serengeti?

To test which of these two theories was correct required a detailed knowledge of food supply, food requirements, and the degree of predation and starvation of the many antelopes—as had been obtained for buffalo and wildebeest (Chapter 8). But for these other species we had no such knowledge; it was too difficult to measure in the early 1980s. However, there was another way to approach this problem. The species could tell us themselves by their behaviour. If the wildebeest were limited by their food supply in the dry season then any other species eating the same grass food would be competing with them. Competition theory predicted that other species such as topi and kongoni should avoid the wildebeest in the dry season when food is limiting but not in the wet season when food was plentiful. In contrast, predation theory predicted no such movement away, and even predicted that species should come together for mutual defence against predators. So, the two theories made different predictions on the behaviour of the antelopes when the wildebeest migration appeared in the dry season in the Mara Reserve. It merely required a detailed record of where each species occurred and what they did in relation to the presence of wildebeest. But we also needed to show that the wildebeest population had levelled out and were limited by their food supply.

* * *

It was 1982. My family and I were based in the Maasai Mara Reserve of Kenya for six months. We were there to count the wildebeest, measure their food supply, autopsy dead animals, and record the movements of all other species in relation to the wildebeest.

The first step was the census of wildebeest as they were on the Serengeti plains in May of that year and so it was necessary to make an expedition into Tanzania for that purpose. Markus Borner, who was based at Rubondo Island in the middle of Lake Victoria, had his plane ready to fly when we arrived. But first we had to find fuel for the plane, and also get ourselves to Serengeti with all our food and petrol from Nairobi in Kenya.

Avgas was of course critical for the count but there was none in Tanzania available to us due to the economic collapse (see Chapter 8). There was plenty of fuel in Kenya but since the border was closed we could not carry it across; exports to Tanzania were banned by the Kenyans, and anyway the Tanzanians would not allow vehicles across the border without some three months of paperwork. Driving trucks loaded with drums illegally across the border was also not an option—they would be noticed. There was nothing for it but to fly the drums in secretly and place them at bush strips in the Serengeti known only to ourselves.

First, therefore, we had to find a pilot and plane. I asked around the flying community in Nairobi with the help of Mike Norton-Griffiths and eventually I found one who would take on the job. I shall call him Emil. He had a Cessna 206, a six-seater which, if we took out the back four seats, could carry two drums of fuel at a time. We needed a minimum of six drums so he had to do this undercover operation three times.

We arranged for the drums to be sent by truck to the Mara Reserve where we were based, and left at the Keekerok airstrip. On the appointed day Emil appeared with his plane. We duly loaded the first two drums into the back of the plane and roped them in as best we could, fully aware that if they got loose we ceased to be a flying machine. The Keekerok staff were very helpful with the loading; they were curious where we were going. To Mt Kenya, we said. This was north of us whereas Serengeti was due south. We took off barely getting into the air by the end of the strip because the air at Keekerok, being at 6,000 feet, was rather thin. Heading north, we disappeared around the hills and promptly turned south.

The eastern boundary of Serengeti is wild country, the ground broken with numerous rocky hills and ravines. No one lived there except remote groups of Maasai pastoralists. Provided we kept away from the immigration post at Bologonja and the Lobo Hotel we were out of sight with a clear run to the edge of the plains where the wildebeest were spread out, and where we planned to cache our drums. It was a good plan and the first run went well, though easing the drums out of the plane with only the two of us at the far end was cumbersome and tiring. The second run went equally well. This was going to be a breeze.

We took off for the third shift, passing the border, the rocky hills far below. A strange smear started to appear on the windscreen, gradually spreading up the window until it effectively covered it all. Emil swore. He pointed to the oilgauge, which was dropping fast. It was only a matter of a short time before we ran out of oil, the engine would seize, and we would be landing like it or not. Emil remembered looking at the oil in the engine after the last run. He had forgotten to put the dipstick back in place, resting it somewhere inside the engine. The oil was being sucked out by the airstream over the engine. We had to land, he announced. I looked out at the chaos of rocks and hills below and pointed out there was no place to land but if he flew about 15 minutes west of us he could land at Lobo Hotel. He could not land there, protested Emil, because he would be arrested and his plane confiscated. Emil calculated that going straight back would only take 15 minutes, the same as Lobo. With that he turned and headed straight for Keekerok, the two of us watching the oilgauge creeping down, anxiety rising in direct proportion. We glided straight in at the strip. We both started breathing again, I think we had stopped somewhere near the border. The airfield staff were surprised to see us so soon, but especially because we had come in from the south. 'What were we doing?', they asked. We distracted them by pointing to the dipstick, still thankfully nestled where Emil had left it. It took a half-hour or so to clean up the engine and windscreen, and pour in more oil. Then off we went again, and this time we had no more emergencies. Our cache was complete, we were ready to do the wildebeest count. All we needed to do was get ourselves there across the border.

* * *

As starts go the one for this expedition to Serengeti in May 1982 could have been better. The ancient bright yellow Land Rover, promptly named 'the Yellow Peril' (and it was to live up to its name), which we had borrowed from Mike Norton-Griffiths in return for renovating it, still looked the same after $2,000, but we hoped it would run better. Four of us were squished into the front seat, Anna and I with our two daughters, Catherine and Alison, in the middle. The back was filled to the roof with everything that we were likely to need because Tanzania had nothing. We were heavily loaded. Setting out from Nairobi we made good progress south and were coming down the long hill towards Namanga, the border post with Tanzania. Suddenly, seemingly out of nowhere a young cow jumped in front of the Land Rover. I slammed on the brakes only to discover there were none. The Land Rover bull bar hit the cow on the head killing it instantly while we continued down the hill at an ever increasing speed. A few miles further down the hill I was able to slow the vehicle by using the gears, and at that moment we arrived miraculously at a garage on the outskirts of Namanga. Pulling in I asked a mechanic if he could look at the brakes and find out what was wrong. After a brief inspection he announced a brake pipe had burst and that he could jerry-rig it so that we could continue; I asked him to do so. At this point a policeman arrived and asked me why I had not stopped at the scene of an accident. I explained that I was unable to stop because my brakes had failed and the mechanic confirmed this. Unimpressed with this answer he repeated the question and I repeated my answer. He explained again that it was my duty to stop at the scene of an accident and, therefore, since I had not stopped he was arresting me. Despite my protestations I was required to report at the police station, which was also the border post. Having fixed the brakes we went on to the border post and I reported. After further similar explanations the senior officer asked me why we were coming to Namanga. I replied that we were going into Tanzania. With that answer he concluded that not only was I leaving the scene of an accident but I was attempting to flee the country. It did not seem to matter that our plans to cross the border had been pre-arranged and that we had voluminous paperwork to prove it. It seemed that they were determined to pin something on me and, therefore, it was clear that I had to do some negotiating. What,

I asked them, was I to do? The two policemen conferred with each other behind the counter and returned to say that I needed to pay compensation for the cow. I enquired how I was to do that, feeling that things were getting out of hand. The lady who owns the cow, the senior officer explained, happened to be right there behind the building. Perhaps I would like to go out and talk with her, he suggested. Whereupon I was introduced to the good lady who explained at length that this happened to be her most prized cow, was extremely valuable, and the breeding potential of her whole herd was in jeopardy. I did not recognize any similarity between this description and the scrawny beast that I had hit. However, I realized that this was all part of the negotiations. I explained that I was leaving the country and therefore did not have much money, and if I gave her all my money I would be left with nothing in Tanzania. After more back and forth we settled on a sum that I could afford and we both returned to the police station where she expressed her satisfaction.

In Arusha we had to buy petrol but none was available, they informed us at the filling station. However, as I got out my Kenyan money there seemed to be a change of heart and we were able to fill up using foreign currency. We also thought to buy some food to supplement what we had smuggled across the border. This proved to be an even more difficult task since there was no food on the shelves of the grocery stores with the exception of a few cans of bamboo shoots from China (we later discovered that the can actually contained one large bamboo shoot which we had to saw up into pieces; it was in any event inedible). We therefore failed to increase our food supply and so pressed on to Lake Manyara where we camped at the rondavels near the entrance. We had intended to reach Ngorongoro but the incident of the cow had delayed us so that we could not get any further before nightfall. We did, however, have a chance to take a short drive around the park; we were the only ones there, since tourists continued to stay away.

Next day we pressed on and arrived at Mary Leakey's camp in Olduvai. There we arranged for her truck to collect the cache of Avgas and food that we had left at the remote airstrip on the plains, and take it on to Ndutu lodge at Lake Lagarja. The Yellow Peril had many disadvantages but one of its great advantages, which in fact saved the expedition, was that it could run on the very low octane—octane 87—of Avgas. And so it was that we had enough fuel to cover our needs.

We camped at Ndutu and met up with Markus Borner and his family. We paid for their stay at the lodge, since he was flying the wildebeest census with me for free in return for my looking after accommodation and fuel. Over the next few days we flew over the wildebeest herds, taking aerial photographs so that we could later count the animals and calculate the size of the population. At the end of three days I asked the lodge manager for the bill, which came to several thousand dollars because of the distorted official exchange rate.

After the census we moved west to the Research Centre at Seronera where we stayed for two weeks. During that time we set about finding the locations where Martin and Osa Johnson had taken their photographs during their three expeditions of 1926, 1928, and 1933 (Chapter 6). This proved to be great entertainment for the family since it was something that they could all do—holding up the original photograph we drove from place to place until we found the exact alignment, indicating the spot where a photo was taken. Sometimes we found a tree that was in the original photo, the same shape of its branches still visible sixty years later.

Eventually it came time to return to Kenya; we loaded up and headed directly north towards the Kenya border and the Mara Reserve where we were based. We got through the Tanzania border post at Bologonja and were on our way to the Kenya border some 20 miles away when with an alarming crunching noise at the front of the vehicle it came to a stop. We discovered on examining the front that the wheels had parted company from the springs and were now resting firmly against the front bumper. It was immediately obvious that the centre bolt holding the wheels and the front springs together had sheared. We jacked up the front, pushed the wheels back over the springs, and let down the jack again. Off we went with the wheels resting on the springs; they were held there by nothing but good faith. We were able to proceed at a painfully slow pace so long as I did not brake. If I did, then the wheels would travel forward as before. Many times we repeated the performance and two hours later we arrived at the Sand River, the Kenya border post. We knew these officials well, and so we decided to camp on the river and sort out the problem later. Next day, leaving the family in camp, I coaxed the decrepit heap the 7 miles to Keekerok where the mechanic was able to replace the centre bolt.

* * *

After another two months it was time to leave the Mara. We were now on holiday, going the long way back around the world to Vancouver, first to Seychelles, then Singapore, New Zealand, and New Hebrides. We headed for Nairobi where we stayed with friends, Richard and Ruth Lloyd. A few days later Richard was to drive us to the airport, early in the morning, for the flight to Seychelles. As we climbed into the car neighbours above our apartment leaned out of the window, and asked where we were going—they sounded incredulous. When we answered 'To the airport', they announced that there had been a coup d'état, and that the airforce had captured the airport.

Hastily we unpacked and went inside again. We remembered hearing rattling in the night; now we knew it was machine guns. There was nothing we could do now but wait to see what transpired. Our house was near President Moi's house. Our daughters entertained themselves in the swimming pool, until the jets screamed overhead and guns started up, which sent them scrambling inside again.

The national radio station, the Voice of Kenya, had been captured by the rebels and they were broadcasting statements that they had taken over the country. Mike Norton-Griffiths lived at Langata on the other side of town and from there he could see the airport. Other friends were staying in the Hilton downtown. We conversed by phone. Military planes were landing at the airport; they had been strafing the President's house. Downtown, looting began, while the police stood and watched. Traffic, bad at the best of times, now reached gridlock and people abandoned their cars in the street. Other cars simply drove off the road wherever they could get around.

A day later things had changed. The army, who had been taking their time to decide which way to jump, came down on the side of the President and moved on the airport. The rebel airforce officers were captured, the radio station changed hands, and the government started broadcasting again. Looting, however, continued in the streets and law and order did not return until the third day, by which time considerable damage had been caused to shop windows along the main streets of downtown Nairobi. Piles of glass littered the pavements, and stores were largely empty.

It was not until the fifth day that we eventually obtained a flight out of Nairobi to the Seychelle islands. Our original plan of a week there had been overtaken by the events in Kenya but we managed to extend our stay another week to explore the islands for their endemic birds—every island had its own unique species. We left for Colombo in Sri Lanka and after a night there arrived at Singapore to be met by Tim, my brother, who was building the Metro under-ground railway there. His first comment was that he did not think there was going to be a coup there. When I looked puzzled at this cryptic comment he told us that Seychelles has just had a *coup d'état* and he had thought we were caught up in it yet again.

* * *

The results of the wildebeest count that year showed that the population had stopped increasing a few years earlier. We had been measuring their food supply, namely young green grass, which was most limiting in the dry season when the animals were in the Mara Reserve. These measures showed that the population was now so large that they were essentially eating out their food supply—it was running out and animals were starving. Consequently the death rate of animals was now balancing the birth rate so the population had stopped increasing. They had levelled out at 1,300,000. More importantly these results showed that the population could regulate itself without anyone having to interfere.[218]

In the Mara Reserve we also observed whether the migrant zebra and the resident antelopes such as topi and kongoni avoided wildebeest or were attracted to them, to test whether competition or predation was allowing all these species to live together. Our measures showed that topi and kongoni actually moved closer to the wildebeest as food became scarcer. This was not what we would have expected if these antelope were competing with wilde-beest for food; they should have moved further away as zebra did. The results suggested that predators were having a far more important effect on the ante-lope numbers than we had previously thought. Predators could be keeping their numbers low enough that they had sufficient food and so were not competing with wildebeest. Instead these antelopes were staying close to wildebeest for protection, using the principle of 'safety in numbers'.[219] This was our first real clue that predators were important in determining numbers of some antelope populations. It changed our view of what was shaping the Serengeti ecosystem. Competition was certainly taking place in large species such as buffalo, and migratory species such as wildebeest, but predation may be shaping the rest of the community.

14

Ivory Poaching

M IKE Norton-Griffiths had proposed that widespread and frequent hot fires were the fundamental cause of the decline in the number of trees in Serengeti during the 1950s and 1960s,[220] as we have seen in Chapter 11. By relocating the sites of the Johnson photographs taken in the 1920s and 1930s,[221] and of others in the 1940s and 1950s, I was able to show that tree numbers had declined for an even longer period, between the 1920s and 1980s.[222]

After the rinderpest had caused the evacuation of people surrounding the Serengeti ecosystem (see Chapter 4) dense thickets of trees and shrubs had grown up, bringing with them the tsetse fly and sleeping sickness disease.[223] This dense vegetation prevailed throughout the period 1900–20 during the German administration (Chapter 5). In 1922 the British Government took over and one of their policies was to reduce the impact of tsetse flies. They did this by instituting a programme of intense burning to open and clear the vegetation.[224] It seems that this policy became ingrained and dry season burning became the norm for decades to come. It had the desired effect of reducing the tree population. So the historical evidence also seemed to support the fire theory.

However, there was one outstanding and inconvenient problem for the new theory of burning. It predicted that with the reduction of burning recorded in the 1970s (because of the increase in grazing by wildebeest) there should have been a mass of young trees appearing. As we have seen, this regeneration did occur in Serengeti although in 1980 it was still too early to see, but in the Mara Reserve of Kenya this regeneration was not happening even when it became obvious in Serengeti.

We knew that extensive savanna had once occurred in Mara from the photos taken by the professional hunter Syd Downey. He had used the Mara area of Kenya as his almost exclusive hunting grounds in the 1930s and 1940s as we have mentioned in Chapter 6.[225] He took clients hunting there—it was wild, remote, and teaming with wildlife, especially with rhino, buffalo, and lions, as well as with wildebeest in the dry season. Elephants were present but not in large numbers. A photograph taken by him in 1944 from one of the characteristic small hillocks on the western side of the Mara River, what is called the Mara Triangle, shows a landscape with so many trees it is difficult to see the grass. So we know that trees are capable of growing there; the open grassland we found in the mid-1980s should have shown the same regeneration that we were seeing in Serengeti because it was experiencing the same degree of grazing from wildebeest and reduced burning. But there was no regeneration.[226] Clearly there was more to this story than we had discovered. This conundrum had to be resolved.

* * *

Holly Dublin set out to find out why trees were not regenerating in the Mara Reserve.[227] By careful documentation of what was destroying small trees, combined with experiments on burning and exclusion of antelope browsers, she was able to show what was happening. It was a surprising result. Mike Norton-Griffiths had already shown for the Serengeti both that fires killed small trees so that they could not replace large trees and that these large trees had died usually of old age and not through elephant damage. Holly confirmed this result for the Mara. However, she discovered a different and remarkable effect of elephants on tree populations; elephants could eat enough small seedlings, even when fires had effectively been prevented by wildebeest grazing, that they could prevent trees from returning and keep the vegetation as a treeless grassland. This was one of the first examples showing how a predator, the elephant, could keep a prey population, the trees, at a low number once that number had been reduced by something else—in this case fire.[228]

* * *

Holly Dublin's result, however, was only half the answer; a second problem remained. Her research explained why there was no regeneration in the Mara. If she was correct then trees could only regenerate when both fires and elephants were absent. Did this situation apply south of the border in Tanzania? In the early 1980s we did not know what was happening to either the elephant population or tree regeneration. I have already described our work setting up the photopoints to measure the changes in tree population (Chapter 11). Now we also needed to conduct a census of elephants and repeat the photographs. Holly conducted an aerial count of elephants in early 1984 in both the Mara Reserve and northern Serengeti; and for good measure she counted the buffalo as well; we had had no information on the buffalo population for the past eight years because of the border problems. Hugh Lamprey, who knew the Serengeti intimately, was the pilot.[229] The result was a surprise—there were no buffalo in northern Serengeti. This area had supported the highest density of buffalo in Africa when we last surveyed it in the mid-1970s. Holly and I examined the results carefully and then reported them to the park wardens of Serengeti.

The authorities in Serengeti were not only surprised with what we told them about the absence of buffalo but disbelieving too. There were plenty of buffalo, they said, they had seen them. Our censuses were wrong. The best way to resolve this, I said, was to do it again, and this time we should cover a larger area. The wardens agreed reluctantly. The date was set for October 1984, with two aircraft, and the whole northern half of Serengeti was to be surveyed. The result was even worse than we had thought—the northern half of Serengeti had no buffalo and the remainder of the survey area had much reduced numbers. Whereas the loss of rhino and elephant due to poaching was indisputable, the same did not necessarily apply to buffalo. Park authorities claimed that a resurgence of rinderpest had caused the drop in buffalo numbers. Poachers, they claimed, could not have done it because wardens would have caught them at it, and also seen the carcases lying around.

One problem was that we still did not know what had happened over the whole ecosystem. So in May of 1986 we carried out an ecosystem-wide survey of all large ungulates. The buffalo results would test the rinderpest hypothesis as opposed to that of poaching: rinderpest should have affected all areas more or less equally so that the decline in numbers would be similar in both north and south of the park; whereas poaching should be more severe in the north and west than in the east and south, which were further away from the villages. The elephant results would tell us whether we had properly understood the interaction between trees, fire, and elephants.

* * *

Stan Boutin, a faculty member at the University of Alberta,[230] and Holly Dublin joined me to help with the counts and the photography of trees. This was Stan's first visit to Serengeti, while Holly had visited only briefly when conducting her elephant work in the Mara. Apart from the aerial surveys we had to climb the hills and set up more 'photopoints' as well as repeat the 1980 photos. To reach these hills required several long journeys across country in our Land Rover.

Alan Root, a well-known film-maker and close friend, had suggested a route west along the Mbalageti Valley from which we could establish photopoints on the central hills of the corridor. These hills were very difficult to reach, little known, and unstudied. Alan offered to come partway with us—it was to take three days to do the trip. I had remembered this area, as did Alan, from the mid-1970s, as open grassland and therefore a good route to follow. We took the usual camping gear, fuel, and equipment in case of breakdown. There was no backup help from the Serengeti Park wardens because they had neither vehicles nor fuel—if anything were to go wrong we would have to fix it ourselves.

We set out in our two vehicles and it was not long before we realized that things were not going to plan. The open grassland that we had been expecting had turned into dense thickets of small whistling thorn acacia trees, only about a yard apart. These acacias have large spines about 1–2 inches long, pointing in all directions, and some of these spines are very strong. In order to get through these thickets, some of which were a mile or more in extent, we had to push through them, knocking the trees over with the bullbar on the front of the Land Rover. The trees were supple so they bent over and then sprang back later. It was inevitable that we were going to get thorns in our tyres, but if the tyres are thick enough this should not matter. However, we had a new vehicle, recently imported from Britain, and it still had the thin six-ply town tyres because there were no other tyres in Tanzania to replace the thin ones.

We drove through the whistling thorn twisting and turning to avoid large trees, stumps, logs, and warthog holes, which can break an axle if the vehicle falls in one; and all the while a cloud of tsetse flies, almost like bees, swarmed around us, biting incessantly. The driver was at their mercy, unable to swat them away while occupied with avoiding the trees. After half an hour of this we were so aggravated by the flies we were glad to hand over to another driver.

The ten or so years since we had been there had altered all the river banks. There had been several years of heavy rains and the floods had washed away the crossing points that we knew about. The numerous gullies had steep banks anywhere from 3 to 10 feet high; they are often dry with sand or pebble bottoms, which are firm enough to drive over if there is a way down an animal path. We had to search for the places where the wildebeest came down to drink. Their constant use over decades erodes the bank to a slope that the vehicle can sometimes negotiate. But the trick was to find a bank where this erosion was present on both sides of the stream, which was not often. At each gully we had to search along the bank, sometimes for a mile or so, with a person perched on top to find a suitable crossing—and this we did some thirty times that day.

By midday we had travelled west a total of 20 miles, though total distance up and down banks and around the bush was much further. We had planned at least 50 miles for the day so we were not doing well. We stopped at a little spring that Alan had visited ten years ago. On that occasion he was looking for sites to do some filming with his wife, Joan, and springs are particularly attractive because there is always lush green grass, bushes, fig trees, and palms, and many birds, monkeys, and other smaller mammals—they are beautiful oases in a sea of thornbush. They had stopped further down the stream and were walking up it, just seeing what was around, when they heard people talking ahead of them. The only people in this part of the park are poachers, illegal hunters. Alan decided to attack them on his own. He crept up and at a prearranged signal with Joan he rushed in screaming, waving, and throwing stones while Joan yelled from the side. The poachers, who had been left in peace in this part of the world for half a decade, were totally unprepared and had the fright of their lives. Thinking they were under attack from a large force of rangers, they all took off up the other bank and vanished into the bush.

A similar incident happened to Alan on the Mbalageti River into which these smaller streams flowed. Again, Alan and Joan were travelling along the bank when they saw vultures dropping from the sky into the river some way ahead. From the Roots' perspective the presence of vultures always means something is dead and potentially interesting things are happening worth filming. So they got out of their vehicle and crept through the bushes. They heard murmuring and squabbling and realized they had come across a gang of poachers just cutting up a buffalo that they had killed—Alan still could not see them, merely hear the low sound of talking. As expected from Alan, he was not going to let them get away with it and he decided to rush in as before. He crept up. Then throwing caution to the wind he leapt over the bank, yelling mightily, right into the middle of them. Joan stayed behind looking over the edge—and what she saw was Alan not landing amongst poachers but on top of a large pride of lions. He was still yelling but the pitch and tone had changed somewhat, and his arms were doing reverse windmill action as in midair he changed his mind about this being a good idea. The lions had never come across anything like this before and being prudent were not about to find out what it was either. With a great roar in unison they all raced up the far bank while Alan was trying his best not to tread on any as he landed and then scrambled up the near bank at the same time.[231]

Alan and Joan left us at the spring to return home, not having any camping gear. Alan promised to fly over us next day to see how we were progressing with this expedition. We pressed on, the thornbush if anything getting thicker, and our progress getting depressingly slower. Towards evening we approached the base of the first hill, Nyamuma, that we wanted to climb—the one we had hoped to reach at noon. We were just in the bottom of a particularly steep gully, examining the bank in front to climb out, when I heard that ominous noise of hissing air. The long expected puncture had happened,

but it was not one tyre: it was three of them simultaneously. We always travelled with two spares but we could not cope with this number of punctures.
A rainstorm was brewing and these small, steep gullies fill rapidly with flood
water. There was no time for us to take tyres off rims and mend punctures,
stuck in the bottom as we were. With some urgency we changed the two
worst tyres and tried pumping up the third to stop it deflating too far. It
started to rain and the bank in front of us began to get slippery—we tried to
get up but we lost traction and merely slid sideways. In desperation I rushed
the bank, the other two pushing, and slowly we eased up, fishtailing as we
went. At the top the third tyre was now quite flat and we had to stop there for
the night.

After supper I decided to show the others how to mend punctures.
I retrieved the puncture repair kit that I always carried with me only to find
that the worst had happened. Whereas there were rubber patches there was
no patch glue in the kit—and with no glue we could fix nothing. We were
now some 50 miles from Seronera, with three wheels flat and a fourth on the
way down, and with no form of contact; it would take at least two days for
one of us to walk back—this being me because I was the only one who knew
the shortest route back.

That night, I pondered over how I was going to get us out of the mess that
I had got everyone into. If we could not inflate our tyres then perhaps we
could put something else into them. I remembered that on the Johnsons'
expeditions in the 1920s they had once, in a similar situation, filled the tyres
of their old model-T Ford with grass. We were at the end of the rains and we
had an abundance of long grass to do the same. But that trick destroys the
tyres fairly rapidly so I decided that had to be the solution of last resort. What
we needed was glue—could we come up with a replacement?

Next morning I suggested that each of us come up with their best suggestion of what sort of glue they could think of. Necessity produces surprising
results for we came up with an encouraging list of possibilities: we carried
small vials of insect repellant against mosquitoes—they do not affect tsetses
unfortunately—and this repellant is notorious for dissolving many substances,
particularly plastics and paint. Perhaps it could soften the rubber to attach
the patch? The tent came with a tube of seam-sealant, which gets very sticky.

Unfortunately this tube was old and had gone rubbery—but by chance we had a small amount of acetone and we wondered whether the sealant could be redissolved.

Alan had promised that he would fly over this day to see how we were getting on, particularly because of this thornbush problem. But there was nowhere to land and we did not have a radio to talk to him, so how could we tell him we were in trouble? I decided to write him a message—in the grass. I had used this trick years ago in Sudan when I used fire ash that looks white against the grass to tell a pilot where our camp was. From the air flattened grass is very visible against standing grass, and spreading what ash we had on top should be better still. I set about dragging a heavy branch to make 10-ft-high letters for the words NEED GLUE—there was enough space between trees to do this—hoping that it would be obvious what it was for. It took me two hours while the others tried the other ideas. In the process of making the letters in the grass, I had to knock down some large perennial herbs with magnificent white flowers that grow in profusion at the end of the rains. The broken stems of these herbs oozed a thick sticky latex, which I got all over me; the significance of this was not lost.

Alan never flew over and so we had to rely on our own resources. We tried all of the possibilities: the insect repellant was a dead loss—it dissolved everything but rubber; the seam sealant was transformed to a glue by the acetone, and the plant latex was also used, although neither of them dried very well. The best and most effective suggestion which I left to last was—sticking plaster (Band-Aids)! These worked admirably: we simply 'band-aided' the patch over the hole. But we had so many holes in the tubes, 30 in all, that we did not have enough Band-Aids and we needed to use all the methods.

I climbed the hill and took the photos. With the view from the top I was able to spy the least difficult route forward and out of this terrible valley. We were ready to leave by mid-afternoon. However, we were now out of patches, so we could not afford another puncture. We went very carefully winding our way through the maze of whistling thorn stands. There was no more bashing down the trees; we went to great lengths to avoid all thorns. I was fully aware that there were probably many more thorns embedded in the thin tyres gradually working their way through to give us the next clutch of holes—it was only a matter of time. We reached Kirawira guard post another 60 miles further west. The rangers had neither vehicle nor repair kits.

We had previously arranged to meet Ken Campbell, who had just started working for the Frankfurt Zoological Society. He had been to Mwanza on Lake Victoria and was returning to rendezvous with us at Kirawira. He should at least have a repair kit—that is, if he showed up, which he didn't. He was new, he did not know his way, and the track across the Ndabaka floodplains was indistinct and surrounded by bogs. As it got dark a ranger appeared on a bicycle to say there was a white man stuck in a swamp about 15 miles west on the floodplain. It could only be Ken. We drove there as fast as we could, praying that we did not get another puncture. We found him at last light, covered in mud from trying to dig out, exhausted from lack of food, and considerably relieved to see us. It turned out that, against advice from the rangers, he had tried to get through at night two days earlier, and in the dark he had lost the track and wandered into the swamp where he had become impossibly stuck. Not knowing the park he was unwilling to walk and so he stayed where he was for yet another night, hoping we would look for him. Little did he know our predicament—we would not have risked searching for him if we had not been told where he was. We pulled him out, took him back to camp, and fed him. Next day we used up all his patches and glue on the two flat tyres that greeted us in the morning, and eventually we got home.

* * *

Our aerial counts in 1986 were designed to give us the information to resolve two problems related first to buffalo and then to elephant. The buffalo counts of 1986 had shown a catastrophic collapse of the population since 1976. Changes had not, however, occurred evenly across the park. Those herds furthest from villages, namely in the south and east, had declined the least, whereas those in the north, next to the truculent Wakuria poachers, had been effectively wiped out. These results supported our predictions that it was poaching and not rinderpest that had caused the drop in buffalo numbers.

In the Mara Reserve Holly had found that despite very low fire frequency young trees were not growing because elephants were preventing regeneration.

We predicted that only when elephants were removed would we see tree regeneration. It was for this reason that we needed both to measure tree regeneration and to census the elephant population in both Serengeti and Mara. The result of the 1986 photography from the hill sites showed that in Serengeti there was rapid regeneration of small trees as we had witnessed on the ground with the whistling thorn. The results of the count showed that elephant numbers in Serengeti south of the border had been reduced since 1980 to only 10 per cent of their original number in the 1970s. In contrast, elephant numbers in the Mara had not changed—if anything they had increased a little. The border was the dividing line because there was a large difference in anti-poaching effort on either side. In Mara there was almost no poaching because there was an efficient anti-poaching field force. In Serengeti there was effectively no anti-poaching and elephants were being killed at a high rate, targeted for their ivory. Elephant numbers had started at around 3,000 animals in 1976 but were a mere 400 by 1986, and these animals were highly traumatized, running at the first disturbance and the sight of a vehicle. Most were cowering on the far eastern short grass plains near Lake Lagarja, a habitat completely unsuitable for them as a source of food but where at least they felt safe—ivory poachers did not go there. Ivory poaching was rampant throughout the years 1978 to 1986; especially during 1982–4 one could hear rifle shots every night, while carcases were found lying by the roads next day with their ivory cut out within sight of park headquarters. So all indicators pointed to poaching not only as the cause for the eradication of rhino, but also for the demise of buffalo and elephant.

In Serengeti, therefore, there was no elephant browsing and small trees were able to regenerate. In the Mara elephant browsing was still heavy and prevented the regeneration. The results were confirming the theory of Mike Norton-Griffiths and Holly Dublin that a combination of fire, wildebeest grazing, and elephant browsing was needed to explain the changes in vegetation, a far more complex interaction than we had imagined.

Elephants could live and feed on trees without reducing tree numbers; the same number of elephants can feed in grassland and prevent trees from regenerating. Consequently we had the interesting situation of elephants keeping an ecosystem in two states—one with a large number of trees and one with a low number of trees. Furthermore, these two states occurred at the same time—one in Mara and one in Serengeti. Elephants cannot move the system from one state to another but they can hold it at the low state, preventing it regenerating and changing into a high state. We call this a 'multiple-state system'. Ecological theory had predicted that such multiple states could exist but at least in the 1980s there was almost no evidence for them, merely a few tantalizing suggestions. Now we had two bits of evidence; just as predators could create two stable states for migrant grazers, as John Fryxell showed for the wildebeest (Chapter 12), so here we have a second case of two stable states with elephants as the predator. Ecosystems can change state from more than one cause.[232]

15

Bandits

THE decline in buffalo numbers had been due to extensive, systematic poaching by gangs of up to fifty people who herded buffalo into snare lines set in the riverine forest. These drives could kill as many as 200 animals at a time. There was a factory-like process of dismembering the carcases, the porters carrying meat out at night on foot as far as 30 miles. The heads were piled under brush and burnt.[233] Poison was set out for predators to stop them eating the carcases.

Law and order hardly existed, the rangers had no vehicles, and it was a free-for-all for the ivory and bushmeat hunters. Ivory was being stockpiled before being smuggled out to Malaysia, Hong Kong, China, and Japan. The bushmeat went to Kenya. All of this was well known to the park authorities, especially the anti-poaching warden Justin Hando. Nevertheless other authorities were still claiming, against the advice of Hando, that rinderpest was the cause of the buffalo collapse. In 1986, after the censuses, a team of virologists and veterinarians—Euan Anderson and Mark Jago from Britain, and Titus Mlengeya from the Arusha Veterinary Centre[234]—arrived to take serum samples from wildebeest and buffalo to test for rinderpest antibodies. Stan Boutin and I helped them collect the samples from buffalo by immobilizing animals. In 1987 we obtained further samples and the results showed conclusively that rinderpest, though previously present, was not the cause of the decline in buffalo numbers.[235] In fact the poaching had occurred almost a decade earlier, between 1978 and 1984. In those years, when there was no food or other commodities in Tanzania, the only place to find them was Kenya, and that required Kenya currency. The Wakuria tribe, who were responsible for most of this poaching, had members living on both sides of the Tanzania–Kenya border. So it was a simple matter for them to carry the meat across the border and sell it for Kenya shillings— monies that could then be used to purchase other essentials. This was the bushmeat trade that was now to affect Serengeti and our research so significantly.

* * *

By 1987 we had some indication from the behaviour of the smaller resident antelopes that predators may be having a much larger influence on their numbers than we had observed for buffalo and wildebeest (see Chapter 13).[236] To test the theory that small antelopes were regulated by predators we needed to find out what was killing them. Large species were easy to find when they died. Their carcases lay around and we were able to conduct autopsies. Small species, however, were dismembered and eaten fast by scavengers such as hyenas, jackals, and vultures, and nothing remained of the carcase within an hour. We knew very little about the causes of their deaths.

However, newly developed techniques were coming to our aid. In particular, radios attached to collars signalled both where an animal was and when it died; we could hear the signal on a receiver and we could detect the direction from which it came. This allowed us to track the animal. The signal was emitted at a particular pulse rate that doubled when the animal was stationary for a certain length of time—a time that indicated it had to be dead. So when we heard this new fast pulse we searched for the radio. From the carcase we could detect whether it had died from predation, disease, or lack of food. Sometimes only the collar remained, but other clues allowed us to tell whether the animal had been killed or simply scavenged. For example, if an animal is killed by predators, it bleeds and a tell-tale blood patch can be seen on the ground. If the animal had died first, before predators had found it, then it would not bleed on being dismembered. Cats, such as leopards or lions, do not eat collars while hyenas sometimes chew on them. The state of the collar gives us a clue as to who was eating the animal. Detective work, therefore, allowed us to build up a picture of what was killing antelopes when we used radio collars. We began by attaching collars to topi and oribi. Later we also put collars on impala.

We built a research laboratory on the banks of the Mara River at Kogatende in the far north-west of Serengeti in 1987. That region was the perfect place to look at the cause of death in several species of resident antelopes when the huge migratory population of wildebeest was not there in the wet season; we would then contrast these results with those when the wildebeest were present in the dry season. Some species were very small such as the oribi, others mid-sized such as impala, while others were large like giraffe. The main scientists conducting this work with me were Peter Arcese with his wife, Gwen Jongejan,[237] and they lived at our camp year-round. Peter was a post-doctoral fellow and in charge of the research. With them was Simon Mduma, who was conducting his Master's thesis under me on the oribi population. We had a camp caretaker, Jumapili, and his wife, Ester. I visited from time to time.

* * *

The signs were there for us to see long before disaster struck if only we had understood them. Early in 1988 sharp rocks were placed on the tracks to puncture ranger (and our) vehicles. A warthog was found hanging in a nearby tree, a defiant message. Later in 1989 we saw people on the other side of the river checking to see whether the river was crossable. Rumours were drifting back to us that villagers thought we were mining gold for they had seen us collecting sand from the river and why else would we be doing that? In fact we were building our laboratory, but no one in the villages could conceive we were looking at animals—we must be there to make money somehow. Fully aware that rustlers were moving cattle across the park (see Chapter 17) we kept away from their now well-worn paths and hiding places in the forest. Nonetheless, we saw footprints on the tracks leading to our camp, bare feet and flat-soled sandals, characteristic of poachers. These people were not just in the park, they were brazenly wandering close to our camp and the ranger post a few hundred yards from us. They were there mainly at night, but not always. Across the river at our camp Peter would occasionally see a person flitting quickly across a clearing in the thicket. We were being watched.

There were other signs of trouble too. The ranger posts of Tabora and Lamai were shot up with machine guns in October 1988, killing rangers and their families. In that month we found a rhino close to camp—the first for fifteen years.[238] We tried to keep this quiet but the rangers soon got to hear of it. A day or two later Peter and I were tracking our collared topi and to get a good signal we were seated high amongst rocks on the edge of a deep valley where we were unseen. From there Peter saw people carrying machine guns tracking the rhino. By the time we had gone back to the ranger post to raise the alarm and then returned with them the poachers had disappeared. The news of the rhino had got to the poachers too fast—someone was tipping them off.

* * *

'On Wednesday November 22nd 1989 at about 7.30 in the evening, just after dark, our Research Camp at Kogatende was attacked by armed bandits.' So began the report from Simon Mduma a few days later. I was in Canada at the time but the rest of the team was at the camp.

Simon continued:

The bandits caused injuries to some of us and looted most of our personal belongings and some research equipment. While the rangers reported to have seen foot prints belonging to three people only, I saw a total of five bandits. Two bandits were armed with guns, a semi-automatic rifle and a .30–06 rifle, while the third was equipped with a bow and several arrows. The remaining two took guard outside in the dark and were never recognized. For the ones who were seen, each had a 'sime' [sword] and one carried a 'rungu' [hard wooden club]. Our camp assistant, Jumapili Shabani and his wife, who are of the Waikoma tribe, identified the bandits' dialect as that of Wakuria tribe.

My house was the first to be attacked. Three of us, Jumapili, Ester and myself had just finished our dinner and Jumapili and Ester were preparing to go to their house located some 30m away. Jumapili was attacked outside and was forced to lie down and remain quiet at gun point. A few minutes later as I was preparing to go out, and not knowing what was happening outside, two bandits entered my house and forced me to lie down and remain quiet by hitting me with a gun butt and rungu. I fell down and surrendered. They later called Ester who was hiding under my [Simon's] bed. They beat her and commanded her to be quiet and obedient to them.

Jumapili and myself were ordered to lie face down on the floor throughout the attack. Serious beatings and harassments on us went on continuously over at least three hours. Our hands were tightly tied to the back with ropes. None of us could move an inch as the ropes cut through our skin causing serious bruises especially at the elbow joints. I lost my upper incisor tooth as a result of being mercilessly hit by a gun butt because I had my head slightly facing sideways. The bandits terrorized us so much that I felt there was no hope for us seeing the November 23rd sunrise.

The bandits were shouting threats, Shoot him dead, one exclaimed, after all he looked as if he was of the Waikoma (there is antagonism between the Waikoma and Wakuria tribes). The other bandit replied that they should not waste bullets, because in time they would cut up the prisoners and throw their bodies into the river [the Mara River, a mere 10 yards from the house]. Another bandit commented that they were not interested in bloodshed. Instead they would merely tie the prisoners' hands together and throw them into the river. The crocodiles would do the rest. Don't think of reporting us to the Police later, one bandit remarked, because we would be dead before that happened.

At that moment I was filled with total despair when one bandit threatened to cut my throat. He claimed that I appeared to be very unhappy with their presence. He placed his 'sime' on the side of my neck and inflicted a superficial cut. I felt the pain. As he was putting more pressure, Ester who was close by, exclaimed in desperation very loudly that the bandit should leave me alone. The bandit left me and went on to threaten Ester that she should not make any noise. At the mercy of God, the bandit ignored me thereafter. As they were satisfied that the three of us were under their control, two bandits with guns went to Dr. Peter Arcese's house.[239]

Peter saw a military type pointing a gun; he was drunk and bleary-eyed. He demanded money and guns—he knew Peter had them for they had seen our dart guns while they had been watching us for months. Peter showed him everything. There was, however, a large camera case in which was kept the money. Peter explained it was full of dangerous chemicals for animals and convinced the bandit not to open it. Gwen and Peter were told to sit in the corner. A second bandit came in with Simon and the others—Simon was bleeding, having had his front teeth knocked out. A third bandit appeared, and there was a discussion in their own language, kikuria, as they ransacked the rondavel.[240] The military bandit ordered the others to tie Peter and Gwen up with animal leather straps, hands behind their backs, sitting cross-legged on the floor. Peter told Gwen to pull her wrists apart as they tied, he did the same, and the bandits did not notice this—luckily they did not understand English. They blindfolded Peter, put him on his knees, and in Swahili threatened to cut his throat—now is the time to be slaughtered, he goaded. Simon pleaded with the bandit not to do it. After looting the house, they left them there, tied up, locking the outside door. Peter commented that they left to discuss, out of earshot, how to kill the five of them because they could not agree how to do it.

Simon continues:

A few minutes later we heard them deflating the tyres of our three cars parked outside. Meanwhile Peter had freed himself from his bonds, because they were loose. He stood up and locked the main door bolts from inside, just in time for seconds later we heard the bandits try to open the door from outside in an attempt to re-enter the house. They found the door bolted from inside. One bandit exclaimed to the others that we had untied ourselves and locked the door from inside. Another bandit rushed to the open window, he tore at the mosquito wire mesh, only to find that the window had a strong metal grill, so that he could not get in.

Luckily we had designed the rondavel to prevent unwanted intruders by using deadbolts on the doors and window bars.

Peter, after bolting the door, broke the glass of a spotlight and using a broken shard cut Simon loose and he in turn cut the bonds of the others. Gwen and Ester then switched off the solar-powered light. They all worked to barricade the door with the fridge, and in case the bandits tried shooting through the window they kept out of the line of fire in the dark, lying flat on the floor. At this point there was a stand-off—they were locked in but the bandits were locked out.

For about 15 minutes after they had found the door bolted we heard them talking and walking outside the house. Thereafter the place was quiet and nothing happened for the next 30 minutes or so. Satisfied that the bandits had left we closed the windows and switched on the lights. It was now after midnight. Dr. Arcese gave us First Aid treatment and we continued to rest. At about 1.30 on November 23rd Dr. Arcese, deciding to take the risk, opened the main door. No one was outside, the bandits had departed on foot, carrying as much as they could.

In the days that followed police and rangers attempted to track the bandits. They found none and all the stolen equipment and goods were lost. Simon and the others went to hospital for treatment; later Simon needed dental work to replace lost teeth. The whole team was in a state of shock.

* * *

For safety reasons we moved our operations out of northern Serengeti; bandits were roaming everywhere, defiantly taunting the rangers at night. The rangers were too intimidated to come out of their houses. Bandits were to run rampant for almost another decade.

Our research on the mortality of resident antelopes continued further south, nearer to the park headquarters, at the old Banagi Hill location for researchers in the 1960s. From the results it was becoming clear that the smaller antelopes such as oribi and impala died from predation, very often by leopards—these predators often hid carcases in bushes, under rocks, or up trees but the radio collars allowed us to find them. Even the larger topi died from predation. But animals as large as buffalo or giraffe were usually too big for predators when they were adults and these died from starvation and disease. Wildebeest were at the transition size. If they were from resident populations, such as in Ngorongoro Crater or western Serengeti, then they suffered high predation, but if they were migrants then they usually died from starvation.[241] Zebra were the anomaly and it took us a lot longer to work out what was happening to them.[242]

16

Of Princes and Polo

THE security situation in Serengeti had reached a crisis by 1989. Rangers were effectively confined to their houses while gangsters roamed at will through the park. The lack of maintenance in the park due to the almost complete lack of operating funds resulted in roads becoming impassable in the rains and river crossings falling into disuse. On the rare occasions that rangers had vehicles they could not access large areas of the park, especially where it counted in the far north-west. There the rivers, lined with forest, were deep and had steep banks. Something had to be done to help the wardens with protecting the park.

Jorie Butler Kent—Mama Kenti as she came to be affectionately known by the Tanzanians—is an American lady of great distinction and influence. It occurred to me that she might just be able to help. Our paths had crossed some years before.

* * *

The phone rang and Jorie Kent introduced herself. She had previously met Holly Dublin, who had suggested she contact me. It was February 1985 and she was at her home in Florida. Some years earlier she had bought Kichwa Tembo tented camp on the Mara River in the Maasai Mara Reserve and consequently had learnt about our research on the Serengeti ecosystem. She had set up a private conservation fund[243] to help both the peoples surrounding the Reserve and the management within it. She wanted to talk with me about what was needed.

She had heard I was coming to New York and was suggesting we should meet. Since I was to give a lecture at the Lincoln Center I could meet her the day before. I asked where I would find her. Just ask for Gracie Mansion, she said, the taxi driver will know. I had to accept this as she rang off.

The New York Zoological Society (NYZS), as it then was called (later becoming the Wildlife Conservation Society), organized an annual meeting for its donors at the Lincoln Center. This was so popular that they filled the Center twice—some three thousand each time—as they held two shows at afternoon and evening sessions. On stage they presented various animals from the zoo and talks on the activities of the Society before holding a dinner at a restaurant in Central Park. They had funded the work in Sudan (Chapter 12) and John Fryxell had now completed his thesis. It was time for him to report on his discoveries concerning the white-eared kob migration in Sudan and they had asked me to talk on the great Serengeti migration.

The following week John and I arrived in New York two days before the great event so that rehearsals could be held. The NYZS representative met me and took me off to the Bronx Zoo, where the NYZS was based, to stay at their guesthouse while John went to stay with friends in Manhattan. Next day I joined John in Manhattan, intending to spend the evening there and then catch a taxi back to the zoo. At 11 p.m. it was snowing hard and we decided to find a taxi to take me back to the Bronx, a journey that went through Harlem. Upon learning where I wanted to go all taxis refused point blank and drove off—apparently they did not like the sound of it.

There was nothing for it but to stay in downtown Manhattan where John was staying. I slept on the couch in my clothes. It snowed all night and by morning there was a thick covering outside. I set out for my appointment with Jorie Kent, dishevelled, unkempt, and unshaven. I told the taxi driver the address. He knew it as predicted. I was curious and asked what it was. It is the Mayor of New York's residence, he told me, down by the East River. We slithered down 2nd Avenue in the slush. Presently he stopped and pointed down a street a foot deep in snow—it had yet to be cleared. 'You will have to walk,' he said, 'I can't get down there.' So in street shoes I walked the 200 yards or so to the entrance, and it was not long before my shoes were filled with melting snow, they were squelching, and my trousers were similarly soaked. The huge coach house doors of the residence had a little porthole through which a porter peered as I pressed the bell. I asked for Mrs Kent, he closed the latch, and presently he opened the doors and let me in, clearly disapproving that I should be allowed in looking more like a hobo than a visitor to the mansion. He showed me to an elevator and sent me on my way. At the appropriate floor the doors opened onto a very small landing, a few feet wide. Just as the doors closed again I realized there was no light and I had not seen any door out. I was suddenly in pitch dark; all I could do was feel for the wall and slowly make my way around until I felt something like a door or a switch—and spread-eagled with arms above my head, face pressed against the wall, was the way Jorie found me when she opened the door. A pool of water was draining out of my shoes onto the floor.

I could hardly have presented a less impressive appearance, but Jorie, ever the gracious lady, welcomed me in to her apartment. Seeing my sodden state she immediately told me to take my trousers off and gave me her husband's dressing gown. Our discussions went well, she told me of her plans, and we went over how she might develop them further. The phone rang. It was the Bronx Zoo, they were in a state of panic. They had lost one of their main speakers for that afternoon—when I had not appeared that morning in the office the staff had asked around and been told that I had gone downtown the night before with the plan of coming back to the Bronx later in the evening. I had not arrived and they feared the worst.

* * *

Shortly after the bandit attack in November 1989 (Chapter 15) I phoned Jorie Kent and presented her with the situation not only of our predicament, but also of the general lack of security in Serengeti. I was confronted with a major problem of rebuilding our lab at Kogatende and of counteracting the bandits. She immediately asked me to the board meeting of her private conservation society called Friends of Conservation, which was timed to coincide with her fund-raising weekend in February 1990, so that she could discuss funding possibilities. A year or so earlier Jorie had asked me to become a scientific advisor to her society. I had already presented her with a plan to build a series of concrete river crossings, or drifts, in northern Serengeti so that rangers could get access to areas they had not been in for a decade or more. She liked the plan. I needed to produce a budget and present it to the board.

This was a special fund-raising event. She had arranged a polo tournament at her estate near Vero Beach, Florida, bringing in the top riders from South America. The commentator who called the play was brought in from Tamworth, Australia. The Patron of Friends of Conservation was also coming. He was Prince Charles, the Prince of Wales, an avid polo player. Guests were to be treated to a two-day tournament of polo followed by a Sunday lunchtime auction to raise funds.

In honour of Prince Charles Jorie held a special dinner at her home for members of the board on the Saturday night. I was billeted with a generous couple, George and Millie Bunnel, at their beach house some 5 miles away, and the society had kindly rented a car for my use—from a worthy company called Rent-a-Wreck. Despite the name their vehicles were perfectly serviceable though of modest make and size. While recuperating from the polo events that Saturday afternoon at the Bunnels' house I realized I did not know where to go for the dinner. It would not do to get lost in the dark on those empty coastal roads, let alone be late for such an important dinner. It would be advisable to do a reconnaissance first. So in beach gear and sandals I hopped into my Rent-a-Wreck and went for a drive. It was easy to see which house was the venue for the dinner. For a quarter-mile before the entrance there were police cars and traffic people stopping everyone. I arrived at the gate and asked if this was the right place to come that evening. I was immediately surrounded by plain-clothes officers belonging to both the British and American security organizations. They demanded to know who I was in not too friendly a fashion. I explained a bit too lightly that I was coming to the dinner for Prince Charles that evening and I needed to see where to come. Since I now knew, I could leave. They had other ideas; in fact they seemed rather threatening. It dawned on me that my appearance was not convincing them I was an invited member of that austere group of guests. After asking for my name they made phone calls to the house a mile or so up the drive and initially the answer was they had never heard of me. They became more threatening as they now noticed my less than impressive car. I was beginning to get worried; I had overlooked this side of the situation. Then the answer from the big house was that they had not received the guest lists yet, so they could not say one way or the other. With that the security people let me go with a clear indication not to come back.

At seven that night I arrived back at the gate with some trepidation, though this time dressed in tuxedo as required. I was again surrounded by security, but this time they were all laughing. Of all the guests I was the one they recognized and they made a grand joke of it, but most of all because I was in a long line of vehicles, all of which were Rolls-Royces, Bentleys, or Cadillacs; the Rent-a-Wreck was clearly different. At the house entrance the vehicles were given over to valets who drove them out to the road again for parking. I asked whether I should show them how to drive it.

Prince Charles, as is the custom, greeted every guest individually as they entered; he had noticed my modest transport approvingly. By an extraordinary set of circumstances an attack by thugs and murderers on a remote camp in Africa had led to a discussion with royalty half a world away on the future of conservation in Serengeti. The irony could not be missed.

* * *

The dinner was particularly important for it raised the awareness of many influential people to the threat to Serengeti as a unique conservation area. The result was a donation from the Friends of Conservation, for both the construction of the bridges in northern Serengeti and the rebuilding of our accommodation there. By the end of 1990 the bridges were complete—one of them was named Mama Kenti drift—and Jorie Kent made a special trip to the north in January 1991, flown in by Markus Borner. She and Markus presided over a parade of the rangers together with David Babu, Director of Tanzania National Parks.

At the same time as these events the Frankfurt Zoological Society under Markus Borner was constructing new houses for the rangers protected by a high concrete wall. They looked rather like the forts of *Beau Geste*,[244] but they gave the rangers a much needed sense of security. These were small steps but they were finally in the right direction. The tide had turned. The war against the bandits was far from over as we were still to see but at least it was a beginning.

17

Hando Fights Back

B Y 1989 wildlife in the Serengeti was in a state of chaos. Buffalo numbers in the northern section, once forming the highest densities on the continent, had been decimated—effectively wiped out. The few remaining elephants had moved out, some to the safety of the Kenya Mara Reserve, others to the far eastern plains where there was little food but at least sanctuary from the killing. Lions in the north were few, removed as the buffalo were snared. Disappointing though these events were, it was our task as biologists to record for posterity the demise of the great Serengeti populations—'there were animals here once', we visualized saying to future generations. So we set in motion the task of monitoring as much of the system as we could.[245] Markus Borner took charge of the aerial surveys of the ungulates with the help of Ken Campbell, who led the Serengeti Ecological Monitoring Program. We continued with our photographic monitoring of the vegetation and the demography of migrant and resident herbivores. Craig Packer maintained his lion studies on the plains[246] and Sarah Durant continued the cheetah studies.[247] Above all we needed to see how the great migratory population of wildebeest was faring. Was this population also collapsing from poaching as were the buffalo? Simon Mduma was given the task of finding out by conducting a detailed study of their births, deaths, and population size, a Herculean endeavour given there were about a million of them covering an area the size of Wales.

* * *

Isengere went by the nickname 'Mwamba', meaning literally a huge boulder but figuratively 'colossus'.[248] He was dismissed from the Field Force Unit, an anti-riot squad in Dar es Salaam, in 1986, an event significant enough in itself since if they could not handle him it was unlikely anyone else could. He returned to his home area, the Kuria village of Machochwe on the edge of north-west Serengeti. During the next two years he gathered a band of henchmen and set out to terrorize the surrounding districts.

Serengeti had until that time been the more-or-less exclusive territory of Maasai cattle raiders for most of the 1900s, and probably since the Maasai had first arrived in the mid-1800s. These Maasai lived in the Loliondo district east of Serengeti, and at frequent intervals during the dry season they would send raiding posses across the park to rustle cattle from the agriculturalists on the western side, there being no one living in the park even in the 1800s[249] (see also Chapter 4). The rustlers would travel the 30 miles on foot by moonlight and then hole up in some convenient thicket within the park near the western boundary at daybreak. The following night they would creep into a village belonging to the Wakuria, Waikoma, or Wasukuma tribes and attack an unfortunate villager, using spears and long knives. Their objective was the cattle held inside small thorn fences, or bomas. Having collected sufficient cattle, and before too many villagers were aroused to fight them off, the Maasai raiders would drive the cattle east across the park. Sometimes they would get the whole way across, but if not they would rest in the park for the day, completing the journey the next night. The irate western villagers would send a retrieving party in hot pursuit but they rarely succeeded in catching up with the Maasai who quickly drove the cattle into Kenya and sold them. The Maasai were hated and feared by the western, agricultural tribes because the few police outposts were quite unable to protect them. Then in 1986 came Mwamba and his gang of Wakuria brigands.

Mwamba set about collecting guns. For the most part guns were only available from government officers licensed to carry them—police, park rangers, and other similar officials. Mwamba told his followers that their 'rite of passage' into the gang was to steal a gun. His rule was simple—he who makes the attack and gets the gun was allowed to keep it. The rest of the gang would provide support in case of retaliation. The first attack came late in 1987 on a lonely outpost of the Game Department in Maswa on the south-west corner of Serengeti. Near dusk, just as a ranger was walking back through the bush to his hut, an arrow swished through the air to hit him. It was a poisoned arrow and it quickly took effect. The ranger died within minutes, silently. By the time his mates found him his gun and killer were gone. At the time the murder was puzzling, an isolated incident far away. No one had heard of Mwamba.

The next attack came in 1988, bold and brazen, taunting authority. It was on the Mugumu police station, in broad daylight, and the bandits used guns. They killed two policemen, making off with the police guns. This was Kuria country: they were attacking their own tribe. Mugumu is a small dusty, sleepy town, one of the main trading and administrative centres for the district. The District Commissioner, the senior administrator, was threatened by the attackers—'You are next', he was told, unless he left town. Quite literally he was being run out of town, and he left terrified, going to Dar es Salaam, the capital. The law of the gun had arrived, it was the Wild West. If the gang was unknown before, it had now achieved magnificent notoriety. Mugumu town became a no-go zone, under the rule of bandits. All those who cooperated with police were warned to get out of town or else.

The third attack was on the prison. On the western boundary of Serengeti resides a prison farm where the less dangerous criminals are allowed to look after dairy cattle. The track through Mugumu into north-west Serengeti runs through the prison grounds. On the west side there is a gate and a sign indicating one is entering the prison. The gate, however, stands on its own, there being no fence or wall either side. The prisoners are always very cheerful, waving to us as they tend the cattle in their white uniforms. Occasionally we would give one of them a lift to the milking sheds, and on one such occasion when asked why the gate was standing alone the inmate volunteered somewhat laconically that the fence wire had been stolen. In any event it was clear that life on the farm with its predictable food and shelter was in many respects more comfortable than a life of freedom and poverty, and the residents were happy to stay. There is, however, a more high security section of the prison with watchtowers. Prison warders man these towers around the clock. One very dark night a gangster stealthily climbed the tower. Next day they found the watchman, dead with a poison arrow in him, his gun gone.

Having accumulated an arsenal of submachine guns and self-loading rifles the Mwamba gang began to use them to terrorize the countryside. In 1988 we saw for the first time attacks and cattle rustling of Maasai villages in the east. There was now a reverse flow of cattle traffic across the north, the cattle being hidden in the riverine forests. The Maasai herdsmen were being shot and killed. Somewhat understandably the Wakuria villagers saw the Mwamba gang and its leader as their saviour—now they could not only protect themselves but they were even taking the war into the enemy's camp. With such sentiments they were not about to tell the police where the gang was hiding.

However, Mwamba & co. were not too particular whom they attacked. Apart from their blatant disregard of government authority they also resented the interference of Serengeti rangers who had the temerity to try preventing the gang from killing elephants and hippo for their ivory, killing buffalo and giraffe for their meat, and running cattle through the park. The rangers needed to be taught a lesson. In October 1988 Tabora guard post was sprayed with bullets, making the huts more like colanders. The rangers cowered on the floors, realizing too late that they had no effective protection. One of them was badly wounded and losing blood fast—he later died of his wounds.

Some weeks later, in early 1989, at Lamai guard post on the Mara River, the Mwamba gang killed the warden, his wife, their newborn baby, and a ranger. Their message was clear enough—the poachers had enjoyed an unimpeded free rein through most of Serengeti for the preceding twelve years of economic collapse, and they saw no reason why they should not continue thus. This was a warning.

If the rangers were lukewarm about patrolling for little pay and few prospects before, they were understandably less than enthusiastic now. Thoroughly intimidated, outgunned, and with no protection, they virtually ceased patrolling in north-west Serengeti.

* * *

The bandit attack on our camp at Kogatende in November 1989 was not an isolated event, of course; it followed from a long period of economic decline.

As far back as October 1985 President Nyerere had resigned in favour of a replacement, President Hassan Mwinyi, who had agreed to the terms of an International Monetary Fund loan. In June 1986, almost overnight the supply of fuel and food recommenced, appearing from the secret stores that had supplied the black market. But the damage had been done. Economic activity took many years to recover; infrastructure such as roads, government buildings, hospitals, schools, and hotels were in a state of decay and needed replacing. Tourism began very slowly to recover but numbers of visitors in 1990 were still only half what they were in 1976; and since these visitors were the main source of revenue to protect the Serengeti from brigands and marauders, the park rangers had little to work with.

A pattern of attacks on tourists developed in the early 1990s as profits from ivory evaporated with the international ban on trade in ivory in 1989. Tourists were becoming more numerous and offered a lucrative alternative. They were loaded with cameras, binoculars, watches, and cash. Even their clothes were valuable. Tourist vehicles that ventured towards the Kenya border, based at the northern hotel called Lobo, were the normal target. A lookout was placed on top of a hill with a view of the road, and this person signalled to the rest of the gang when a suitable vehicle hove into sight. They usually had a good ten minutes to prepare. Tourist vehicles were easy to identify for they did not resemble the dark green Land Rover pickups used by rangers. Rocks or a tree were placed across the road around a corner so the driver did not see the impediment until it was too late. The vehicle would stop and become surrounded immediately, the gang threatening to shoot the driver if he moved. The hapless tourists, terrified, would be ordered out, stripped to their underclothes with shoes removed and told to walk away from the vehicle, after giving up all their possessions. The driver would sometimes be clubbed and so disabled. Then the bandits would make off through the bush. Eventually the driver, if he was lucky, would be able to drive to the nearest ranger post or hotel for help, usually several hours after the gang had vanished. By the time the rangers were on the scene it was many hours or even a day later, and although rangers were excellent trackers the trail was lost.

Our research vehicles were easy to identify: they were glaringly white, hardtop, Land Rovers or Toyotas. We were easy targets. We were forced to adopt an anti-guerrilla strategy to travel to the north of the park when we were monitoring wildebeest in the dry season—we being Anna and I, and variously Simon Mduma, Ray Hilborn, and his wife, Ulrike. We knew the hills that bandits used as lookouts so we approached them cautiously and studied them with telescopes, keeping our vehicles in dead ground and out of sight. If all seemed clear we sped through the 10 miles or so at top speed to prevent the gang from setting up the road block. At the same time we took to employing a ranger on our bush work, something we had never done before. This ranger was stationed on top of the vehicle within full view, his rifle at the ready. He was quite literally 'riding shotgun'; this was indeed the lawless Wild West. If there was a ranger vehicle travelling in the same direction we would go in convoy, and with much hilarity several rangers would sit atop our vehicle spoiling for a fight.

Sometimes they got their wish. The various gates to the park took in tourist fees, and these were collected by an accountant in a National Parks Land Rover each month guarded heavily by armed rangers. On one occasion in 1992 they had to travel to the north-eastern gate called Bologonja, right through the most dangerous area for bandits. Unfortunately there was no Parks' Land Rover available so the accountant borrowed the tiny Suzuki that barely held four people—used for tourism by the warden, who rarely ventured out in it. It was a type often used by tourists. Into this little sardine tin squeezed the accountant, driver and, in the back, two rangers and rifles. The car made it safely to the gate, collected the cash, and set out on return along the hills in easy view. Before long they were met by the thugs on the road, who ordered all to get out. The front people got out hands in the air, followed with considerable difficulty by the rangers who had to climb through the front seats—and they immediately started firing. In no time the gang was put to flight, several were wounded, and all were captured. This story made the rounds of the ranger posts for months with considerable mirth. It was a great morale booster at a time when the bandits had the upper hand.

* * *

It was January 1993 and I was staying at our research house with Simon Mduma, his wife, and their small baby. Simon was in the middle of his thesis work on the wildebeest population changes. It was a wild and stormy night, heavy rain pelting onto the metal roof made conversation difficult, winds were driving the rain horizontal, and lightning and thunder were continuous. It was not a night to be out. Sometime after going to sleep, towards midnight, a heavy banging on my door awoke me. It was Simon shouting over the din to wake me up; he was telling me that bandits were attacking the house next door, used by the lion researchers. Pulling on my clothes and shoes as I ran out I asked what was happening. He explained that a girl, Ester, who had been with us when the bandits had attacked at Kogatende three years earlier (Chapter 15), had been sleeping in the house to guard it. The researchers[250] were away in Arusha and had left Ester in charge. She had heard the bandits breaking windows and climbing in the front, assuming no one would hear on a night like this. Ester, with remarkable coolness and presence of mind, climbed out of a back window and ran over to our house some 100 yards away to alert us. Jumapili, our caretaker, Simon, and I, climbed into the Toyota and slithered on mud-slick roads, the rain coming down in sheets, round to the lion house. There in the constant flickering blue lightning, deafening thunder, and drenching rain we saw people, naked except for loin cloths, and carrying bows with quivers of poison arrows over their shoulders, running with lots of household goods on their heads—mattresses, beds, clothing, kitchenware—anything they could get their hands on. The scene was surreal as we three, unarmed, roared in, horn blaring trying to make as much noise as we could and appear more numerous than we were. We raced towards them shouting, waving wildly, but making very sure we got nowhere near those arrows. The bandits, encumbered as they were, could not see who was approaching and so they dropped their loads, even their bows, and raced for the river. It would have made the Keystone Kops proud.[251] Soon they were all gone and we were faced with securing the house, collecting as much property as we could see in the rain and lightning, and taking stock.

Simon, pleased that this time he had got the better of the villains (after his horrendous experience three years earlier at Kogatende, Chapter 15), drove into Seronera with his wife, who refused to be left alone now, to inform the police post there. I remained in the lion house for the night in case the bandits chose to return, something we considered unlikely so long as they could see there was someone in the house—I left a lantern burning for this purpose. Next day we set about recovering as much of the goods as we could find, lying scattered for hundreds of yards across the landscape. (We heard later that we found virtually all of their belongings). We also found bows, quivers, and arrows, which we collected and displayed as our booty of war—not for long, however, for the police impounded them as evidence, much to our disappointment.

It was not all good news, however. As the 1990s progressed, the bandits became more daring, expanding their activities from the northern areas not just to the centre where park headquarters and our Serengeti Wildlife Research Centre were situated, but also to the southern parts. On the road to Sopa Lodge in the Mbalageti Valley they hijacked two tourist vehicles and killed an American tourist. In 1994 a researcher was attacked one night in his house, his wife and child held captive until he produced the monthly wages he was keeping for his employees. By now, the mid-1990s, the security situation was not only intolerable, it was desperate. Tourists were being killed and the news was getting back to the West; the tourist industry and newly recovering economy of Tanzania were threatened. Something drastic had to be done.

* * *

Justin Hando is a quiet, modest man if you meet him in a social situation. A man of the Iraqw tribe, a group originally from Somalia and unrelated to surrounding tribes in the Mbulu area near Ngorongoro. Soft-spoken, almost shy, he has a disarming smile. He is likeable, approachable certainly, but not the self-assertive commander of men one would expect of a person in his position—or so it would seem on first meeting. We met in 1986 when he first arrived as the warden in charge of anti-poaching, second-in-command of Serengeti. It was a phantom force then with no vehicles and little money. David Babu, by then the Director of National Parks, came to Serengeti in June of that year and called in to see me—we had been friends since 1968 when he was a stripling trainee warden. We talked and the outcome was that all of the scientists were persuaded to lend their vehicles, themselves as drivers, for a massive anti-poaching exercise. Loaded with rangers under Hando's command, the vehicles patrolled either side of the Orangi River from Banagi to Musabi in the western corridor. Poachers, taken by surprise for they had not seen a vehicle for almost a decade, ran in all directions. Rangers took after them in hot pursuit but failed to catch any of them—unfit, untrained, and ill-equipped they were no match for the bush-hardened poachers. But they did come home with piles of snares, dried meat, and poison arrows loaded into the backs of the research cars. This was a first strike back at the hunters who had defiantly displayed their scorn at the ineptitude of the rangers for years. It may have been a token operation but it was something.

David Babu went to great lengths to raise funds from international donors, the Frankfurt Zoological Society being the leading contributor amongst several others, to purchase vehicles, modern weapons to replace the ancient .303 rifles, uniforms, and other equipment. Over the next three years a well-trained field force began to take shape under Hando, but they were still considerably outnumbered by the gangs of bandits. In October 1988 a ranger force in Land Rovers was patrolling along the Mara River forests and Hando was with them in the front seat. Machine guns stuttered, bullets flew through the front window; the driver took evasive action and drove for safety. By a miracle none in the front was hit though Hando received glass in his eye, and a ranger standing at the back was wounded. They came to our camp at Koga-tende and we arranged for a plane to be flown in to take the wounded to hospital. We talked with Hando—we saw a completely different person, the opposite to his public persona. Standing beside the vehicle he exuded a silent anger, a relentless determination—he was going to get his revenge. He was not going to stop until he got everyone of those murderous bandits.

In 1993 Hando was promoted to Chief Park Warden of Mikumi, a park near Dar es Salaam, but the mayhem continued as outlaws rampaged through the villages and the Serengeti for the next four years. Not all of the villages were prey, of course, for some of the Wakuria tribe were providing succour and shelter to the gangsters—and the most notorious village was Machochwe, from which many of them came.

By 1996, with a state of anarchy prevailing in northern districts, Hando, now in charge of security for all of the Tanzanian Parks at the headquarters, was very quietly transferred back to Serengeti into his old position of anti-poaching park warden—on the surface a puzzling move since it appeared a demotion. Hando was his old unassuming quiet self: he would come to visit us, spend long evenings with us, and find out what we had seen around the Serengeti since we were out there in the remotest corners every day, he almost deliberately not commenting on issues to do with poachers. Patrolling was not obviously any more frequent and we wondered why they had brought him back.

Something was different, however, at first almost indiscernible, like a small stir of air, but then as time passed it became a clearer pattern, a real breeze. Tourist vehicles were still being ambushed with depressing regularity, and we still had to carry armed guards and could not camp except at ranger posts. However, whenever tourists were attacked, the rangers got onto the trail with unlikely speed—often within half an hour. The posse tracked the bandits for hours, never giving up, three rangers tracking, two ahead as scouts to avoid ambush, and Hando was often with them. On their return they had with them the stolen property and increasingly the guns that the highwaymen carried.

Bandits do not leave their guns in the bush to be found, nor do they give them up willingly. Firefights were taking place and one by one there were fewer gangsters at large. Lobo, in the east of the park, was attacked by a gang of ten bandits; they were carrying seven rifles. The three rangers were outnumbered so could do little but follow the gang while informing Hando. They found the camp on the Orangi River and returned to Lobo to meet Hando, who had organized a posse. Hando arrived at midnight and they had just one night-viewing glass. First they drove, then at 3 a.m. they walked for two hours across the bush, stopping to scan from high ground. Finally near dawn they saw a light, just one small firebrand, but it was enough to tell them where the gangsters were. The gang had moved east and were now resting on the side of a large termite mound surrounded by open ground, some on each side so that they could see anyone approaching. They were now 70 yards away. Hando's men quietly took up position; one of them was sent to one side as a decoy. The decoy put up his head and made himself visible; it had the desired effect. Instinctively all the bandits came round to look and started shooting. But they were now exposed and an easy target. There was no contest when Hando's rangers returned fire. A few bandits tried to run but not for long.

After a while it became evident that the rangers knew a thing or two somewhat in advance of events—were they getting inside information, we asked ourselves. As 1996 passed into 1997 there were fewer attacks though they still occurred, and the gang was being led by an authoritative person wearing a beret—clearly Mwamba. But the rangers were fighting back and villagers began to notice when no one came back.

The showdown took place in February 1997. It became known that Mwamba and what remained of his gang were hiding in the forests near Wogakuria. The police at Mugumu were told. They arranged for Mwamba's sister to walk the 20 miles or so across the forests and rolling hills to the gang and tell him to give himself up. They sent her back with the answer that they would kill her if she reappeared, and understandably she refused to do any more missions. The police sent in a unit of several vehicles. They knew the location; the gang was in a forest strip along a small river some 200 yards down the slope from the track the police were using. The police commander stood on his vehicle to be easily seen, and using a loudspeaker he demanded the gang give themselves up, they were trapped, and there was no hope for them. Two shots rang out. The police bodyguard and driver fell dead. The rest of the force beat a hasty retreat back to Mugumu to assess the catastrophe.

Hando sent in his rangers—by now secretly trained special commandos, crack shots, and expert in tracking. They followed the gang relentlessly, day after day, and one by one there were fewer bandits on the run.[252] Eventually there was only Mwamba left—and he had disappeared completely. Or so it seemed.

The Wakuria live on both sides of the Tanzania–Kenya border. Mwamba had slipped across the border and was hiding in the tribal villages of Kenya— someone had given a tip-off. The Kenya police were asked to help and they started to search, which served to flush Mwamba back into Tanzania, though no one knew where exactly.

Not long afterwards a member of the special field force, one of Hando's agents, dressed as a local villager, was having a beer in a bar at the notorious village of Machochwe. There he recognized Mwamba. The ranger engaged him in conversation, offering him a drink, and while the unsuspecting Mwamba had his hands on the bar the ranger deftly handcuffed him. Thereafter, events moved swiftly as the ranger informed his boss in Seronera, and the police at Mugumu. The police were only too happy to have this criminal in their hands and soon had him behind bars. All that remained was to wait for the trial at which he was surely to be convicted.

Or maybe not. After a few weeks the inhabitants of the Wakuria villages started to raise funds to be used for his defence and release, he being a hero in their eyes. The police, realizing that such a tactic may well be successful, decided to conduct further investigations, ostensibly to improve their case of murder. They escorted Mwamba back to the scene of his most recent killing at Wogakuria Hill in Serengeti. When they returned to Mugumu with Mwamba he was found to be mortally injured—he had been wounded as he had tried to escape, or so they said. Whatever the case, Mwamba died of his injuries a few days later.

So ended Mwamba's eleven long years of terror, murder, and mayhem. It had set back Serengeti conservation by preventing protection of animals, the economy by deterring tourism, and our monitoring by reducing our ability to gather information. After the dust had settled we saw how Hando had done it. His quiet, self-effacing manner had deluded the enemy into thinking he was a pushover. Meanwhile he had very quietly and secretly set up a network of informers, no easy matter since the Wakuria hated all other tribes. So Hando had had to use members of the same tribe who spoke the same language—quite literally money talked. There were six of these special agents, each in a different village. Gradually he had accumulated the information that allowed his team to be at the right place at the right time when attacks were made. At the same time, he trained a team of rangers who were experts in the bush as trackers and as crack shots. Equipped finally with modern weapons they were now a match for the bandits with their AK-47s.

Hando remained for a year longer in this position. In August 1998 the then Chief Park Warden, Maregese, died suddenly from blood poisoning. Hando was instated as Chief Park Warden and remained thus until 2006; we were the best of friends. He was arguably the best CPW that Serengeti has ever had to that time.[253]

* * *

The Frankfurt Zoological Society under Markus Borner played an integral role in these events by supplying the funds and infrastructure. Without Markus the bandits could not have been beaten. Our monitoring showed that after 1998 the ecosystem began to recover from the long decades of poaching and pillaging. Simon Mduma's research showed that wildebeest numbers had been able to withstand the onslaught of killing and there had been no decline between 1977 and 1999 with the exception of a natural event, the great drought of the twentieth century in 1993 when we lost about a quarter of the population, which I recount in Chapter 19. Wildebeest numbers returned to normal after only three years. Elephant numbers increased dramatically; nearly every mother had a baby in tow and they were giving birth at their maximum rate. By 2007 they were almost back to where they were in 1977. Buffalo did not fare so well because meat poaching continued in the north and numbers there were so low that they could not get going. But elsewhere in the system they started slowly to increase and in areas furthest from people they had almost returned to normal by 2007. Only black rhino remained at low numbers. At a minuscule 10 animals in 1998, down from their original 500 in 1977, they were so few that simple accidents of life were sufficient to slow their increase; over some fifteen years they had increased to about 30 animals by 2012. Clearly the only way to speed up the return of rhino to the ecosystem is to reintroduce new individuals to inflate numbers to a point where they could overcome accidents and show a sustained increase. With the support of the Frankfurt Zoological Society and the American hunter and conservationist Paul Tudor Jones reintroductions began in 2010.[254]

In 1997 we were finally able to return to documenting the Serengeti ecosystem in peace.

18

Man-Eaters

THE great rinderpest epidemic of 1890 killed much of the prey of lions so rapidly that lions were left stranded with nothing to eat. It is no surprise, therefore, that starving lions turned to eating domestic animals and even humans: outbreaks of lion attacks on humans occurred most famously in the Tsavo area of Kenya. It was at the village of Tsavo in 1898 that Indian labourers brought in to construct the Mombasa–Nairobi railway were the target of man-eating lions—some thirty people were killed. John Patterson's famous book, *The Maneaters of Tsavo*, recounts his efforts to hunt down and remove the two lions supposedly causing the killings.[255] Attacks also occurred in Uganda, and in Sukumaland just west of Serengeti. Indeed, such attacks contributed to the evacuation of agricultural land south-west of Serengeti at the beginning of the twentieth century, in concert with the increase of tsetse flies and smallpox, all brought on by the rinderpest.[256]

Although lion attacks on humans declined in later decades as wild ungulate numbers increased again, such attacks continued on a low level in populated areas of East Africa. In recent decades lion attacks on humans have become more prominent as the human population increases exponentially at 3 per cent per year. Areas that once supported wild ungulates have now been cleared for agriculture, the wild animals hunted for food until little remains. Once again lions are left with no food, and once again they have started to feed on cattle and humans. Predation on domestic animals occurs outside of protected areas such as Tsavo in Kenya,[257] Tarangire,[258] and the Ngorongoro component of the Serengeti ecosystem in Tanzania.[259] Lions tend to feed on cattle while hyenas and leopards feed on sheep and goats. Lion predation on humans is particularly prevalent in southern Tanzania. Craig Packer[260] and his associates Dennis Ikanda, Bernard Kissui, and Hadas Kushnir, have investigated the causes of these attacks.[261]

Man-eating lions must be removed and this is the task of the Tanzania Game Department. Simon Mduma, before he became a scientist, was such a Game Department officer in southern Tanzania during the 1970s and 1980s. Part of his job was the hunting of man-eating lions.

* * *

We were camped in the corridor on the Grumeti River just downstream from Kirawira guard post, Simon and I, in August 1994. We had been running his counts of wildebeest carcasses, doing autopsies on dead animals, and looking for predators. Now it was dark, we had eaten, and were sitting around a camp-fire. It was a very still night, very dark, not even starlight, and silent, no zebra yelping, not even a scops owl. We talked quietly. There was the faintest of cracks, a twig breaking. Simon shone a flashlight, and there in the beam was a lioness crouched, tail switching, not 10 yards away staring at us. She had been stalking us. We kept the beam on her and presently she crept away still watching us; we could see her eyes shining in the beam. To be sure she was at a more comfortable distance we followed her until she was half a mile from camp. We returned and sat around the fire, conscious of the lion out there in the dark, and the still night seemed threatening. Had he told me about man-eaters, Simon asked me, he had had to hunt them once. He proceeded to tell me his story; it seemed the right occasion.

* * *

It was 1978, and Simon Mduma, only twenty years old, was a very young Game Officer with five askaris (rangers) under him. A ranger post south of Mahenge in the Selous Game Reserve of southern Tanzania, where Simon was posted, reported that a lion had become locked in a house while searching for victims. He and his askaris went to the village and found that the lion had entered an open door, which had then closed itself behind the lion. The problem was how to get the lion out. They drove up to the door in a vehicle, and leaning out, tried to push the door open, but the door had locked itself. Kazimoto, the senior askari, volunteered to walk to the door and open it. As the door opened the lion leapt out and jumped on Kazimoto. He raised his arm, which the lion grabbed, and they both fell to the ground, one arm in the lion's mouth. Kazimoto wrapped his legs tight around the belly of the lion, preventing it from doing more damage but consequently the others could not shoot. After a while the lion paused, and taking this opportunity another askari with a long knife walked up to the lion and eviscerated him. The lion jumped up with Kazimoto still attached. A third askari then managed to shoot and kill the lion, but Kazimoto was so frozen with fear that he could not let go. It took some time for him to recover his senses and then he began to shake uncontrollably, experiencing post-traumatic shock. He kept his arm, badly scarred.

* * *

Simon became the District Game Warden at Liwale in southern Tanzania in 1980. In that area a man-eating leopard had learned to climb through grass thatch roofs. The leopard would seize a victim, and while everyone else ran away in panic the leopard would drag its victim out through the roof again. One day he received a report of this leopard eating a victim, still in the house. Simon and his askaris arrived and tried to disturb the leopard by throwing rocks into the house. The leopard, however, stayed inside where it felt safer. One of the askaris then opened the window to look in and the leopard came out of that window so fast that no one had time to fire a shot. The spectators, quite a crowd who had gathered to watch events, turned and ran in panic in all directions. The leopard reached a group of three and jumped them from behind. He scalped one, who died, and injured the two others. Then the leopard simply disappeared in the confusion of the crowd running every which way.

They decided to trap the leopard using a goat inside a very heavy trap. However, the leopard was very canny; it sensed a trap and would not go in. It simply walked around the trap, departed, and went to another village where it killed another person, as always through the grass roof. It did this every few days in different villages.

One of the curious circumstances of these events was that all the victims were related. There developed a strongly held belief associated with this family and the super-leopard. The story goes that a young man had his wife stolen by another man. He went to a hill where a spirit lived—a spirit called Mahoko. The man talked to the spirit and explained his plight. Mahoko asked what he wanted. He wanted his wife back. So Mahoko sent the leopard to attack the family of the man who had stolen the wife.

The villagers told Simon not to bother with catching the leopard, it was protected and under the control of the spirit. Instead they pleaded with Simon to tell the man to return the wife. Simon ignored these fanciful stories and continued trying to catch the leopard. One day his askaris, having set the trap with the unfortunate goat, returned to the village to wait. As they approached in the dark they saw a leopard in the middle of the road. An askari dismounted from the vehicle and shot the leopard; they were much surprised by the behaviour of the leopard, waiting to be shot. They loaded the carcase into the truck and returned to Liwale some 30 miles away. Next day there was much rejoicing in the village and Simon was told why—the bride had been returned. The villagers interpreted the events to mean that the leopard was a message— it was no longer protected, its job was done.

Intrigued by stories of this spirit Simon asked to see the Mahoko, who it was said lived on a mountain. It was arranged, therefore, that one day he was to be taken there by villagers. The first part was by road, but then on foot for several miles. Chickens were everywhere wild in the bush, a curious and unusual situation. At one point they removed their shoes at a sacred place, a place where they could smell lions. Here he was required to leave his gun. They walked another half-mile where they found lions completely undisturbed by their presence. He stopped and his escorts went on; he was not allowed to go further into the sacred place, it was as far as the villagers would let him go. On foot, unarmed, and with lions around him, his ten minutes there felt like an hour. The escorts returned after conversing with the spirit and they all walked back to their vehicle. Simon was nonplussed; he was not sure whether he had been hoodwinked or whether this sacred place had lions that had become habituated to people since it was taboo to disturb them. However, there was no doubt that the people believed in Mahoko.

PLATE 19 President Nyerere (3rd from right) meeting the scientists, 1973, introduced by the Director of the Institute Dr Mcharo (centre).

PLATE 20 Collecting data on wildebeest deaths 1972; the silver-backed jackal could not wait.

PLATE 21
Mike Norton Griffiths examining
the drowned wildebeest at
Lake Lagarja, 1973.

PLATE 22
Anna had to follow our hot air
balloon 20 miles across the plains
while Catherine kept lookout.

PLATE 23
Counting elephant using the Supercub at Tsavo National Park, Kenya,
1972.

PLATE 24
Mary Leakey (left), Mike Norton
Griffiths and the tame cheetah,
Olduvai 1977.

PLATE 25 Hiking the southern hills. Anna recording photopoints, 1990.

PLATE 26 The increase in Acacia woodland density. Photos from the same point on Kimerishi hill (a) 1980 (b) 1986 (c) 1991 (d) 2011.

PLATE 27 White eared kob at Boma National Park, South Sudan, 1982.   (Photo by J. Fryxell)

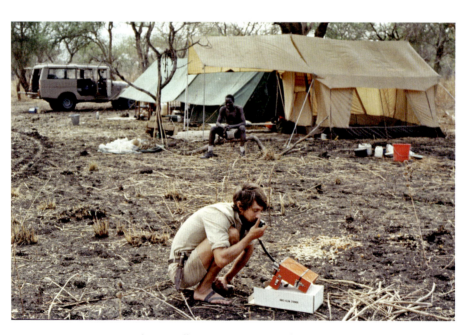

PLATE 28 John Fryxall in camp at Boma Park, South Sudan, 1982.

PLATE 29 Murili hunters at Boma were happy to watch us all day and night.

PLATE 30 Predators cannot migrate with the wildebeest because they have to carry their young.

PLATE 31
Elephant maintain open grassland
by weeding out seedling trees in
Mara Reserve.

PLATE 32
Kogatende camp on the Mara river
before it was attacked, 1989.

PLATE 33
Hando's driver and the bullet holes
after their ambush 1988.

(Photo Peter Arcese)

PLATE 34
Rangers accompanied us on our
research journeys, early 1990's.

PLATE 35
Markus Borner and FZS kept the
Serengeti going through the 1980's
and 1990's.

(Photo courtesy Fromman/Laif)

PLATE 36
Simon Mduma autopsying wildebeest
1993.

PLATE 37
Male buffalo in the drought of 1993
were too weak to fight off lions.

PLATE 38
Migrant animals in the drought
of 1993 risked their lives with the
crocodiles to get water.

PLATE 39 Wildebeest shape the whole ecosystem. An exclosure shows that wildebeest grazing maintains the short grass plains, 1986.

PLATE 40 The Mara River is the lifeline for the Great Migration of wildebeest and zebra: irrigation and deforestation in Kenya are threatening the water flow.

* * *

It was 1983 and the place was Liwale in southern Tanzania. Simon was by then a 26-year-old, experienced District Game Officer. He had received a call from the village, very early in the morning, by messenger. A young mother had been killed by a lion not far from her house while she was fetching water. Neighbours had been roused at dawn hearing her chilling screams as she died. They rushed to her rescue but the mother was already dead.

The messenger had cycled 20 miles to get him. Simon talked to his assistant Mohammed Mtila, who cycled around the town to alert the rest of his askaris. In haste they climbed into a Land Rover with shotguns and rifles; there were four of them in all. In due course they reached the corpse; and the villagers said that the lion was still around. They were in an old cashew nut farm with small stunted trees and thickets. They decided to climb the trees in anticipation that the lion would return although the trees were not very tall. Once they were up the trees they had a clear view of the body. Then the waiting began—for two hours nothing happened; it was now mid-morning. An old man walked along the path and saw the four of them clinging to the top of the small trees. He scoffed at them—perched, he derided, like scared monkeys. He could not believe what he saw—game scouts with guns sitting in trees. He thought they were all wimps. He pondered aloud that no wonder they were all being killed by wild animals.

An askari shouted to warn him of the lion. He brushed aside the warning, saying he would follow the path and run through the bushes. At that moment, from nowhere the lion jumped clean over the thickets, landed on the man, broke his neck, then jumped back again out of sight. He was so quick that not a single guard moved, let alone fired a shot. Now they had a second body on the ground lying with his face turned 180°. All were stunned, struck silent for five long minutes. No one could believe what they had seen. They knew the lion was still there and that the lion had jumped higher than all of them in their little trees—the lion was too fast for them and all were clearly easy prey. They knew it and no one felt safe.

The lion was still in the thicket, they had heard him snarling. No one wanted to get down from the trees to be the next victim. Simon asked himself what to do next. Some of the askaris fired randomly until they realized that if the lion ran out they would be worse off not knowing where it was. Also, Simon realized that if they injured the lion it would be even more dangerous. So they waited again for another three hours past noon, as the sun became hotter and hotter, still precariously balanced in the tops of the trees. Finally they plucked up enough courage to climb down, the furthest ones first, until all converged in a group on the two bodies on the path. They retreated backwards, guns at the ready, keeping an eye on the thickets.

The villagers were less than impressed with this performance—the outcome from the arrival of these so-called game scouts was that there were now two bodies instead of one, and the lion was still at large. Simon had to do something. First, they decided to remove the bodies and to do this they had to clear the thicket. Rounding up by now reluctant and scornful villagers they advanced through the thicket with a line of beaters hoping to flush the lion out. No lion appeared. They realized the lion was still in there watching them, capable of jumping anyone at any time it felt like it.

As dusk approached the askaris had to find food and shelter in the village although they were embarrassed to accept hospitality after what had happened. They explained shamefacedly to the elders what had gone wrong. Clearly to regain the confidence of the village they needed to demonstrate they had tried. Next day at dawn they went back to the thicket, climbed the trees, and waited. Again after a period of time they got down from the trees and threw stones into the thicket, but nothing appeared. Finally, becoming a little braver, they went in and saw tracks leading out, only then realizing that there was not one but *three* man-eating lions—at least the lions were not there but now no one knew where they were. The lions were still hungry for they had not eaten either kill, and so they were likely to strike again. Things did not look particularly reassuring.

They tracked the lions through the bushes, knowing that the lions could probably see them well before they would see the lions. Quietly one askari signalled. All knelt, they saw three lions in the bushes. Simon called up three of his sharpshooters. They took aim and at a signal fired simultaneously. Two lions were killed immediately but the third was wounded and ran off, which meant that they had to track it. However, after twenty minutes they managed to cut it off and kill it.

With relief they returned to the village to tell them the good news. The villagers came out to carry the lions back; there was much dancing and singing in the village that night. Some ate lion meat.

* * *

Tunduru is the district headquarters of south Selous, the famous game reserve of southern Tanzania. It was 1984 and Jaboma was the Game Warden for the district. Lions used to wander through the town hunting domestic dogs while leopards went for the chickens. There were reports of man-eaters in the neighbouring village. Jaboma went to this village with four askaris, he himself carrying his .38 revolver. Having dropped off the askaris at the village to deal with the problem, Jaboma returned to Tunduru and parked his car at his office. By this time it was dark. He walked in the dark across the small airstrip near his office to his home. At the end of the airstrip was a ditch. Jaboma was jumped there; his body was found the next day. His wife thought he was with the askaris in the next village and so did not raise the alarm when he did not return. The body was almost entirely eaten; his hands and the revolver were all that remained to identify him.

Simon was travelling that day to Tunduru to pay the staff salaries and was told the story when he arrived. It was the same lion that they had been chasing in the next village 10 miles away. The villagers all felt that the lion knew Jaboma was after him and had decided to get him instead.

Simon became convinced that it was time to move on and find another life; it was only a matter of time before it would be his turn. He went to Dar and enrolled in the University, which is how he came to be a scientist.

It was late now, time to turn in. We shone the flashlight around and were not particularly reassured when we saw nothing. We were both a bit on edge. There is something sinister about man-eaters.

* * *

Craig Packer and his team have shown up some of the causes for lions becoming man-eaters.[262] Incidents of human predation by lions have increased dramatically since 1990, especially in southern Tanzania where wildlife has been reduced. The human population effectively doubled between 1990 and 2005 so that more land has been transformed into agriculture. The lion researchers found that in some areas of Tanzania the main problem for agriculturalists is the bush pig, a secretive animal that feeds nocturnally and is surprisingly common—it can live in heavily cultivated areas with dense human populations provided it has a refuge, usually a swamp or wet thicket. Bush pigs love to eat beans and other legumes. Peasant farmers protect their crops at night by building a small shelter with a bed raised off the ground—the shelter is open on all sides so the farmer can see his crop and chase off the bush pigs. Lions like to eat bush pigs and follow them out to the crops, and in this situation they come across farmers sleeping in their shelters. This presents too tempting an opportunity for the lions, and some have taken advantage of it. Soon the lions become accustomed to eating humans. Packer has advised that the best solution for avoiding the human–lion conflict is to control bush pig numbers, if we want to conserve lion populations. Tanzania has one of the largest lion populations remaining in Africa.

Of particular importance to conservation in the Serengeti ecosystem is the retaliatory killing by Maasai of lions and other carnivores when these kill livestock.[263] In the Ngorongoro Conservation Area resident lions kill cattle and these lions become the victims of Maasai warriors. In addition, there is ritual killing of nomadic lions that move from the Serengeti National Park. Hyenas are also subject to retaliatory killing. A more ominous trend has appeared in the 2000s with the use of poison to kill carnivores. Although poison is banned, it is easily available and becoming more frequently used around the ecosystem. Unfortunately vultures are highly sensitive to such poisons and large numbers of deaths are now being reported. The concern is that vultures will disappear in Africa as they have in India.[264]

19

Biodiversity

I N 1992 the nations of the world met in Rio de Janeiro and set up the Convention on the Conservation of Biodiversity.[265] Tanzania was a signatory and as a result was committed to documenting and protecting its biota. The Serengeti supports a highly diverse fauna but apart from the mammals very little of it had been properly documented. I began to look at this biodiversity. What species occurred in Serengeti? Why was it so diverse? What caused changes in the species living there? Did the diversity play a part in how the system worked? There is considerable diversity in Serengeti of both large herbivore species and predator species; did such diversity contribute to the stability of the system? Scientists were then debating whether diversity was helping to maintain and stabilize natural ecosystems.[266] If the presence of many species helped to stabilize an ecosystem then the loss of species might result in a collapse of that system. Clearly this issue had significance not only for conservation but also for the persistence of human systems because when humans change natural habitats to agriculture they reduce the natural diversity by a very large amount.

The wildebeest were the key to understanding how the Serengeti ecosystem worked. By the mid-1990s we had become aware of the tremendous influence the migrants had on most parts of the Serengeti ecosystem. This led me to ask: to what extent did the wildebeest population maintain or change the biodiversity of Serengeti? If wildebeest numbers were much reduced—as they had been following the great rinderpest epidemic—then would this change in population affect the biodiversity of the system?

In addition, our understanding of the ecology of large mammals in Serengeti, particularly of niche partitioning, facilitation, and regulation through food or predators, all depended on whether the migratory wildebeest population was limited by food supplies in the dry season. We had by the early 1990s strong circumstantial evidence that wildebeest were limited by food similar to our evidence for the regulation of the African buffalo population (Chapter 8). Such evidence was based on direct measurements of both the food supply and the death rate as wildebeest numbers increased and levelled out in the 1970s and 1980s.[267] However, we needed to test this theory. In principle, one way to do this would be to reduce the food supply and measure what happened to the population. We predicted that more animals should starve, causing the population to decline; subsequently numbers should increase when food supplies returned to normal. If the population did not decline or increase again then we would know that some other influence such as predators, disease, or human hunting was controlling events. Of course we could not do this reduction of food as a planned experiment—it was after all a national park where we could not interfere, and anyway the scale was vastly greater than anything we could achieve. But a natural experiment presented itself when a major drought occurred in 1993. This drought was more severe than any that had so far been measured in the twentieth century; it was just what we needed to see whether we were right in our understanding of the system. Simon Mduma set out to measure the population responses to the drought. Ray Hilborn, a world-renowned computer modeller of populations, joined us for six months.[268]

To understand changes in biodiversity and their consequences on the ecosystem I set up a monitoring programme to record the biodiversity of Serengeti. In particular we studied the effects of climate, predation, grazing by the migrants, and wildfires. The main research focused on the large mammals but we also looked at other parts of the system including birds, rodents, insects, and plants.[269]

* * *

It was late in 1993 that events started to become extreme. The drought was so severe that hippos no longer had water to wallow in and many were lying out in the sun, dying slowly of heatstroke. Crocodiles had retreated to the last patches of mud, catfish slithering over their backs as they crammed on top of each other to stay wet. But they had plenty to eat as zebra and wildebeest, crazy with thirst, threw caution to the wind and marched into the few remaining water pools. Buffalo were standing around too weak to move while lions were running up to them and knocking them over—two, three, four at a time—sometimes not even bothering to kill them let alone eat them. Wildebeest were moving out of the park and into the villages in a desperate attempt to find water and green food. Villagers were chasing them down the main street, killing them with spears.

We were all overwhelmed with the number of dead animals that we had to autopsy. We were looking for evidence of starvation of course, but also for the effects of disease, and predation. Many of the carcases were in an advanced state of decay by the time we reached them so it took a hardened stomach to cut into them and look at their bone marrow.

Ray and his family made a remarkable discovery. They saw and filmed the deaths of lions in early 1994.[270] This was the beginning of an epidemic that killed some 40 per cent of the Serengeti lions in that year. Craig Packer, Sarah Cleaveland, Andy Dobson, and others went on to establish that the deaths were caused by canine distemper. Until then no one had known that this disease, typical of domestic dogs, could kill lions.[271] It also seemed to be affecting hyenas. It came at the end of the great drought, suggesting that contact between wild carnivores and domestic dogs could have been greater at that time when wildebeest were out in the villages. This was the first time canine distemper had been found in any wild cats.

* * *

Over the years we had lions around our house often; at night they delighted in sitting on our sofa that we left on the verandah, and cubs would play with the cushions, dragging them between their legs as they would drag a wildebeest in later years. In the morning we would search around in the long grass to retrieve our cushions, now with large holes in them—those cushions have been recovered many times. The lions became accustomed to our presence, sometimes visiting our birdbath to take a drink as we sat on the verandah at night not 5 yards away. We even had an incident when lions attempted to kill a male buffalo who took refuge in our carport—he would not leave, knowing it was safer there than outside, and caused us considerable problems for a few hours.

Our normal trapping for rodents and insects took place twice a year in four areas of the ecosystem—at the end of the short rains in January and long rains in July. In January 2002 we went west to the Ndabaka floodplains, an area of flooded grassland with stands of a silvery-coloured, very spiny *Acacia seyal*. These stands have been known to me for a long time, and by now they were mature, a forest of silvery-red spiny and spindly trees, all of one species, one of our special habitats. I had first visited this site thirty-seven years earlier in 1965 with Arthur Cain (Chapter 7), when it was a dense thicket of young trees, so thick that we had to crawl along tunnels made by animals. It was while we were on our hands and knees, with me in the lead, that I met a leopard coming the other way. It was so narrow that we could not turn around (the spines prevented this), nor could the leopard, and an uncomfortable stand-off resulted—accompanied by much snarling from the leopard. I did not feel too safe but there was nothing much to be done except back out the 100 yards and that took ages—well, it seemed to me rather too long. The leopard did likewise.

Now we were to carry out a general biodiversity survey, recording as much as possible on the birds, rodents, reptiles, and insects over three days. We caught high numbers of rats and shrews, every trap was filled, and sometimes there were two or even three rodents in a trap. There were many black-shouldered kites and long-crested hawk eagles around, a sure sign that rats were in high numbers.

We camped at Kirawira, using Richard Bell's old rondavel to cook, eat, and store our food. This was as well since there were some fifteen vervet monkeys around, all of whom had learnt to steal our food. All we could do was throw stones at them but these they just dodged—in the morning they descended on us like a pack of thieves and raced through the door, almost between our legs, to grab whatever they could find; they were so fast we could never catch them. I had pitched my tent in the riverine forest along the Grumeti River about 50 yards away from the rondavels under a great tamarind tree, sandy,

shady, and by a huge pool in the river containing the biggest African crocodiles on the continent. The river was full from all the rain we had received
upstream, and the hippos were swimming back and forth all night, calling
and splashing. Needless to say Simon, Nkwabi, and Stephen thought I was
mad—if the floods did not wash me into the river then the crocs would come
and get me, or hippos would drag my tent (plus me) into the river; the end
result they claimed would be the same, namely croc food. I pointed out to the
team that I had camped here many times and it was not subject to flooding or
hippo highways and crocs don't come and get you. Hippos did indeed walk
off with Tim Corfield's tent when he camped here with myself and Vesey in
August 1965, and I recounted this story to them as a matter of interest. That
settled the matter as far as they were concerned; they were not about to test
my hypotheses about croc bait or hippo trails. So they camped way out in the
scorching sun.

That first evening, while we were sitting around the campfire, Makacha
told us the story of a ranger at this Kirawira guard post, some five years earlier, who insisted on washing in the river by standing on some rocks in the
middle. He liked these because the water was cleaner than that at the muddy
banks. The others warned him that he was asking for trouble, but he dismissed them, saying the rocks would protect him and he would see the crocs
coming (though what he would do about it remains a puzzle). All they heard
was a shout, he was never seen again, not even a scrap. These crocs are so
huge they could swallow a man in one gulp. Now the rangers have a borehole
and pump their water for washing and drinking.

We had lions around camp every night; on the first one there was a fight
between territorial males that lasted most of the night, the roaring making
sleep impossible. The lions chose the team's campsite as the battlefield; it was
a tense night. In the morning we found the lions on a wildebeest kill just outside camp and not far from the ranger post. One of the lions suddenly got up
and started stalking through the bushes. Looking to see what she was after I
saw to my astonishment an old man walking directly towards the kill—the
lioness was stalking the man. With Simon driving we raced over to get
between the lioness and the man. As we came closer I recognized him as
Kisiri, the uncle of our assistant Jumapili. He had heard we were in the vicinity

and had decided to walk over from his village and see us. He had lived his whole life surrounded by wild animals and was not in the least bit concerned about walking alone through the bush; but this time his luck was about to run out. He was quite surprised as we roared up to him and fairly hauled him into the vehicle; he thought we were making entirely too much of it.

* * *

Simon Mduma completed the study of the wildebeest population in 1999, showing conclusively that numbers were set by the amount of green grass food available in the dry season. The great drought of 1993 had provided the critical evidence that animals starved in large numbers. In the two years that followed, the birth rate was very low because many females failed to conceive and many newborn babies died from poor lactation. Then things returned to normal and numbers increased again back to where they had been, in the region of 1.3 million.[272] This was the answer we needed in order to understand the rest of the changes in the Serengeti region.

Craig Packer's long record of data on the lion population combined with previous data by George Schaller, Brian Bertram, Jeanette Hanby, and David Bygott, covering some forty years, showed that lion numbers increased following the wildebeest increase. Lion population increases went in jumps; prides first increased in size as food supply—the wildebeest—increased, but when they reached a critical size the pride split and occupied a larger area with two territories.[273]

The increase in wildebeest numbers followed by that of lions also provided the clue to some other anomalies that had puzzled us. Early records from the period 1900–30 had indicated that both roan antelope and wild dog were widespread in the Serengeti savanna country (Chapters 5, 6); by the 1990s both had effectively disappeared from the ecosystem, occurring only at its edge.[274] Perhaps the low numbers of wildebeest in the early twentieth century had also resulted in low numbers of lions. We have evidence from the work of Norman Owen-Smith, Gus Mills, and colleagues in Kruger Park, South Africa, that when wildebeest numbers increased so also did those of lions, and roan antelope numbers dropped almost to extinction due to increased predation.[275] Similarly, wild dog numbers in Serengeti declined progressively from the 1960s, when we first started recording numbers, until they disappeared from the savanna country in 1992. Part of this decline was due to higher numbers of hyenas and lions that both stole the kills made by the dogs and killed the wild dog pups. So both roan and wild dog populations may have declined as populations of the top predators—lions and hyenas—increased after wildebeest rebounded from the rinderpest epidemic. Roan and wild dogs may both have been able to exist in the system only as temporary residents while wildebeest were in low numbers—they benefitted indirectly from the great rinderpest epidemic.

Grant Hopcraft used Craig Packer's data of lion kills to show that lions were most successful in capturing their prey when they ambushed them at suitable sites along river gullies, even if the prey were avoiding such sites. Most of the antelopes needed water and so at some point they had to go to the rivers. Lions did not go to where the main herds were feeding because they could not ambush prey there.[276]

* * *

By the 2000s we had accumulated evidence on the factors that limited numbers of many different species of ungulates, some very small such as oribi and Thomson's gazelle, and others very large such as buffalo and giraffe. At the same time we had recorded what species of predators had been killing these ungulates. When all of these data were put together we saw an exciting pattern emerge: large ungulates had very few predator species eating them; indeed very large ones like elephant had none when they were adult. In contrast, very small ungulates such as oribi were prey to as many as seven species of carnivores. This happened because large predators had a much wider range of prey sizes than small predators—lions can eat prey from the 1,000-lb buffalo to the 10-lb dikdik, whereas the small serval cat can only eat small ungulates, hares, and rodents. So a small dikdik is eaten by many carnivores, and as a result virtually all of them are killed by predators. Dikdik or oribi numbers are limited by predators; they never reach levels where they run out of food. Large ungulates suffer very little predation and most of them die from lack of food and disease that kills them when they are in poor condition.

The conclusion is that both food and predators limit ungulate numbers but which one operates depends on the size of the species. More importantly this pattern depends on the diversity of both prey and predator species: many prey species are required to support the many predator species (but not as may appear on a one-for-one basis), and many predator species are required to limit small prey species. The twin process of predator limitation of small prey, which we call top-down control, and food limitation of large prey, which is bottom-up control, results in a stable system of predators and prey. Diversity was indeed important in shaping this complex pattern.[277]

* * *

The riverine forests along the Mara River are slowly disappearing. Greg Sharam, who was studying these forests, showed that disturbances such as wildfires burning into the thickets and elephant browsing opened up gaps in the canopy. The openings allowed tall grass to grow; grass could not survive under the canopy. Forests with open canopy lost many of the fruit-eating birds so that tree fruits fell to the forest floor uneaten. Such fruits suffered very high infestations of weevils and were killed, and the remainder did not thrive in competition with grass. Under closed canopy the seeds that were regurgitated by birds suffered very low weevil attack. So regeneration of the forest declined when there was disturbance and the few seedlings were insufficient to replace the adult trees. In other words a stable forest required the fruit-eating birds to provide sufficient seedlings; once the birds were gone the forest declined. Again the presence of many species of birds provided stability to the system.[278]

Disturbance also affected the bird community in the grasslands. Nkwabi showed that when either fire or wildebeest grazing changed long grass into a short sward different bird species appeared; these were adapted to short grass. The wildebeest migration maintains the short grass plains: we know this from studies where animals were excluded from patches of short grass plains for many years and a community of long grass species took over. Mary Leakey told me that when she started work on the archaeology of Olduvai Gorge in the 1930s the plains near there were long grass; this was a time when the wildebeest population was less than 10 per cent of what it is now. The

plains in that area are now short grass. So wildebeest are effectively supporting a suite of bird species in the Serengeti ecosystem.[279]

Other studies have shown that when agriculture changes natural savanna
habitats at least half of the insect- and seed-eating bird species are lost, and
there is a decline of some 80 per cent in the numbers of the remaining species.[280] These massive losses that occur with agriculture are also seen in many
other areas of the world and they emphasize the need to maintain areas of
natural habitat large enough for bird populations to survive.

* * *

By the end of the 2000s there was one more big question that had been
troubling us: what maintains the stability of biological communities in the face of
disturbances from weather? We had by now developed a picture of how the
community of species, whether plants, birds, or mammals, influenced the
stability of the ecosystem and how disturbances such as burning could cause
changes in that system. But there was one aspect that we had not yet explained.
Individual species populations are stabilized by feedback mechanisms affecting births and deaths, as we have seen in the Serengeti buffalo and wildebeest.
But communities are composed of thousands of species and if they all operated independently of each other there would be a gradual change as some
species drop out and others come in—the relative proportion of species
would change markedly over time. Yet this does not appear to be happening;
over the past 4 million years the number of carnivore species in the Serengeti
ecosystem has remained between 9 and 19; in the last 2 million it has remained
between 9 and 14 species.[281] The types of species have evolved with new
forms replacing old forms of similar ecology so that the number remains
relatively constant. Over the past fifty years we see that the relative population sizes of the grazing mammals have remained the same—there is always
about 1.3 million wildebeest (after they had recovered from disturbance) and
8,000 kongoni, some 200 times fewer in number but very similar in ecology. This could occur if the species are all reflecting the relative ratios of
their resources or niches. But we now know that many, if not most, of the
species are regulated by top-down processes where resources are not limiting.[282] So just as there are regulatory mechanisms for individual populations,

we suspect there are similar mechanisms for whole communities. We already had a few indicators: John Fryxell showed that predators hunting in groups, like lions, on prey that lived in herds (most of the large mammals) resulted in a very stable predator–prey system; it rarely got out of balance.[283] Similarly the complex community of predators where small species feed on a range of the same prey species narrower than that of the larger predators (food niches of smaller predators lie within the niche of larger species), which we have already mentioned above, also leads to stability.[284]

The first clue we had as to how this mechanism might work came counter-intuitively not from observations of stability but from instability: we had noticed that parts of the ecosystem fluctuated quite noticeably for no obvious reason. For example, rodents exhibited plagues every few years but we could not explain why these occurred. Wildebeest calf and yearling numbers also fluctuated, roughly in phase with the rodents, which was even more puzzling. Then in 2010 we found the answer and also an even more significant discovery.

The temperature on the surface of the sea north of Australia influences the direction of ocean currents, which in turn affect climate around the world. The climate of Serengeti is affected by the surface temperatures of these seas, usually known as the El Nino Southern Oscillation. The temperature is recorded as variations from an average sea surface value, and such variations above (positive) and below (negative) the average are known as the Southern Oscillation Index (SOI). We have found that the variations in survival of wildebeest yearlings, topi yearlings, rodent outbreaks, area of savanna burnt each year, and even disease outbreaks are all related to the SOI. This occurs via rainfall, which is the climatic factor that responds to SOI. More importantly, different parts of the Serengeti complex respond in opposite ways to the SOI so that when one part suffers, another part benefits. For example, when SOI is positive there is more rain in the dry season and wildebeest yearlings survive better, which in turn provides more food for lions and they survive better. However, these years also result in less rain in the short rains, which means that rodents survive less well and so do the small carnivores and birds of prey that depend on them. The opposite processes occur when SOI is negative.

As a result the system swings according to changes in SOI. At first sight this would mean more rather than less disturbance in the system.

However, some parts of the system benefit from both extremes of the SOI and this double benefit results in stability of the system. For example, zebra yearlings benefit when the SOI is both positive and negative. Predators that rely on wildebeest yearlings would be at a disadvantage if such yearlings were few in number (in fact when SOI is negative) but they may be able to supplement their food by switching to zebra. A stable predator population can then create stability in the ungulates since most of the smaller resident ungulates are predator-limited.[285]

So we find that the diversity of plants and animals plays a role in how the whole ecosystem responds to disturbances from climate. We now must see whether other parts of the world respond in the same way. For Serengeti, however, it is clear that the presence of many species together explains the patterns in the ecology of the system. Loss of this diversity will affect how the Serengeti operates.

20

The Future of Conservation

In accepting the trusteeship of our wildlife we solemnly declare that we will do everything in our power to make sure that our children's grand-children will be able to enjoy this rich and precious inheritance.

The conservation of wildlife and wild places calls for specialist knowledge, trained manpower and money, and we look to other nations to co-operate with us in this important task—the success or failure of which not only affects the continent of Africa but the rest of the world as well.

—Julius K. Nyerere, First President of Tanzania and
Father of the Nation, *Arusha Manifesto* (1961)

E ARLY in 2010 the Serengeti came under the greatest threat to its existence since it was formed in 1950. The Tanzanian Government announced that there would be a major trunk road through the northern part of Serengeti National Park. There followed an international outcry against an action that was likely to threaten the survival of the ecosystem.[286] But this is not just any system; it is universally seen as one of the great wonders of the natural world, and the world was making that statement.

The threat had already been on the horizon since the national election in 2005 but it remained in the background until the next election loomed in late 2009. Letters asking the Tanzanian Government to reconsider were sent from many international organizations including the International Union for the Conservation of Nature (IUCN), the Worldwide Fund for Nature, and several governments. Serengeti is a World Heritage Site and the responsibility of the United Nations under UNESCO; this group also wrote to the President.

An important dilemma arose amongst conservationists. Should they maintain a low profile and try to persuade the Tanzanian Government behind the scenes? This has the advantage of not embarrassing the government and so allowing it to change policy. Or should conservationists make a public stand, creating an international issue to show that this is a world asset? The former approach may have been the better one tactically, but it had the disadvantage of creating a bad precedent: it would leave the fate of a world heritage site at the mercy of whimsical decisions, ones that may well have unwanted outcomes through lack of information. This approach also gave the wrong message: no international outcry implied a lack of concern; it was acceptable to compromise and threaten the Serengeti and, therefore, encourage road development. The latter approach was more risky for it could make the government intransigent, but it had the important advantage of publicizing the scientific facts and creating the precedent that no one has the right to do as they wish with unique world assets. Some conservationists opted for the former policy; the majority for the latter.

Scientists put together the biological facts derived from the forty years of accumulated information.[287] The proposed road was to cut the northern extension of the park in two. This is the area to which the wildebeest and zebra migrate in the dry season as their refuge when food and water are in short supply. They head for the Mara River, the only permanent water supply in the ecosystem, and the region of highest rainfall where they can find green food. Migrants remain in this region from June to November, the duration depending on the vagaries of the rainfall; they usually move outside the park boundaries on both the west and east sides, just as they did 100 years ago although now not so far.[288] Numbers of animals are in the range of 1 million wildebeest, zebra, eland, and gazelle. In addition, large herds of buffalo, elephant, and topi are resident there. All of these animals could be feeding near the road, crossing it, and because it is open and flat, resting there to ruminate.

There are many roads in the Serengeti National Park but none of them produce the human–wildlife conflict that the 'north road' is likely to create. The present roads are made of gravel, locally called murram, and are used by local traffic for tourism, a few buses, and local trucks. These trucks are small because all traffic currently passes up the tortuous, steep, and narrow Ngorongoro escarpment. This road is impassable for large 18-wheel trucks and semi-trailers; indeed they are not allowed to use the road. The traffic cannot travel at more than 50 mph.

The proposed road has all the problems that we have seen with other roads through wildlife areas around the world, only on a much larger scale. The problem with such roads is not that they impede wildlife—they do not—but that they encourage fast driving. With such a vast number of animals on the road there are bound to be accidents, just as we have seen at Banff National Park in Canada with moose and elk—and that was with only 800 elk. In Mikumi National Park in southern Tanzania, where a road was built through the middle in 1972, the greatest number of deaths from road accidents occurred in species not disturbed by traffic; these species in Serengeti are the migrants. There is now a growing body of evidence that roads cause increased human access for poaching, wildlife deaths, isolation and fragmentation of populations, and both loss of and disturbed habitat. In general roads are a major cause of biodiversity loss.[289]

Accidents cause human fatalities, as occurred at Banff, and soon enough there is pressure to build fences along the road. These fences, of course, will prevent the migrants reaching their precious water and food, and so there will be a collapse in numbers. Fences in southern Africa cause problems for wild species, natural communities, and whole ecosystems; indeed they even increase human–wildlife conflict.[290] The calculations have been made for the Serengeti ecosystem.[291] Everything we have now discovered in the Serengeti ecosystem shows that it is entirely dependent on the huge populations of migrants to keep it the way we know it. Without the migrants it will change into a different system; the Serengeti as the last remaining major migration on the continent will be lost.

Major roads attract settlement along them and so bring habitation to the very edge of the park boundary. Inevitably, fences will be constructed along the boundary. The problem with fences is that wildebeest and zebra do not know what they are. They have never met them in their lives and are not adapted to tolerate them. They usually do not see the wires, or they think they can push through the fence; and so the great herds run straight into the wires.

Fences were built across the short grass plains at Angata Kiti in the Gol Mountains in 1964 to keep the wildebeest out. When the wildebeest arrived they stampeded and the fences fell over within a few minutes.[292] Fences built to withstand the charging masses will result in catastrophic mortality with bodies lined up along them. Lions will use the fence to trap prey, as they have with the fence around Keekerok airfield in the Mara reserve. A major human–wildlife conflict would develop where none now exists.

* * *

As a result of the information from long-term research and the arguments we have outlined, the Tanzanian Government announced in June 2011 that it would not go ahead with the north road as previously planned. Instead the government declared it would confine the tarmac to roads east and west of the park, and consider a southern alternative if the international community would fund it.[293]

The reason for a road south of the Serengeti ecosystem is that the peoples in the Mara Region west of the park can achieve greater benefits from development. Although a road cutting through the Serengeti would be shorter than that passing south of it by about 50 miles, the cost of building that road would be more expensive because it would have to climb the Rift Valley escarpment along a new route, whereas the route south is already established. Also, the southern route provides more economic returns for Tanzania because it services almost twice as many people, and three times more agriculture and livestock than the northern route. Because most of the northern road traverses uninhabited areas, these economic spin-offs would not be realized for the Tanzanian economy. In general, the southern route provides access to the Mara Region using many roads that are already built and which merely need upgrading.[294]

Currently the government intends to confine the tarmac to roads east and west of the park. In between, across the park, there would remain the current gravel road to be used only for tourists. Meanwhile, the government has asked for help in finding funds for the southern route around the park. This is a positive step because if the southern route is built, then there is no economic need for the northern road. However, the future of

Serengeti depends on international funds being found for construction of the south road.[295]

* * *

The northern trunk road across the park is not the only threat to Serengeti. Development outside the ecosystem has impacts inside it. There are two serious threats from outside. First the Mara River is the most important water source for the migration in the dry season. The animals move north to the Mara River because it is the only flowing river of any size in the dry season and provides vital water for the millions of animals. Unfortunately the Mara has its origins in the Mau forests of the highlands of Kenya. These forests are being cut down at an accelerating pace and the flow of water has declined accordingly. There is perhaps some hope here because the Government of Kenya is developing a restoration programme for these forests.[296] However, these plans may be undermined by uncontrolled water offtake downstream. The Mara River flows through agricultural land where unregulated irrigation is tapping off the water upstream of the Serengeti ecosystem. As a result the Mara has in the late 2000s stopped flowing during the dry season for the first time in a century: these impacts take place in Kenya, out of reach of controls from Tanzania.[297] Unless agreements are drawn up to regulate the irrigation offtake, particularly in the dry season, the wildebeest could find there is no water for them when they arrive in the next few years.

Secondly, Lake Victoria is a vast but shallow lake immediately to the west of Serengeti. It is, as we noted earlier, so large that it creates its own weather system, producing rainstorms in the dry season in the west and north-west of the ecosystem. It is these storms that provide the food for the wildebeest migration. The lake has only one major water inlet, the Kagera River on the west side, and some smaller rivers on the east side, the main one being the Mara River. If the Mara stops flowing there is a possibility that the lake, being only a few feet deep around its shores, could recede as water levels drop, and in many places recede several miles from the current shores. Millions of people dependent on the lake for fishing, irrigation, and drinking water would find themselves stranded. In addition, the weather systems of the lake could be changed, with dry season rain declining and thus changing the migration pattern of the wildebeest. The drying up of the Aral Sea in Asia due to uncontrolled irrigation offtake should be the cautionary tale for Lake Victoria.

* * *

The Serengeti north road, the drying of the Mara River, and the retreat of Lake Victoria threaten both a unique world asset and the most important source of foreign income for Tanzania—Serengeti attracts the majority of tourists and its loss would affect the economy of the country. There are lessons we can draw from these threats for the future of conservation.

Conservation has been approached from two directions. Because much of the natural resources, including biological species, lie in areas used by humans there has been a push towards trying to conserve species through sustainable use or at least limited impact. This approach is known as community-based conservation (CBC). It is an essential approach because humans now cover over 90 per cent of the terrestrial surface, which means that about 50 per cent of the world's species lie within this area. In developing countries CBC has focused on exploiting natural areas to provide income for local peoples in the hope that they would value the areas and the native species that lie within them.

On the surface this idea sounds convincing but in practice it has not been able to overcome several intractable problems. The fundamental problem is that the approach is short-sighted.[298] By definition all CBC areas have people living in them, and these people expect to receive their share of the natural resources—for example, they could receive the income from selling a portion of the harvest of wildlife in an African game reserve. Such a plan may be acceptable when first devised, the harvest having been correctly calculated for the numbers of wild animals that live in the area and the income shared out fairly. Projects of this sort are advertised as great successes for CBC.[299] However, before very long things go wrong. First, the number of people increases, as it does everywhere in the world, and there is a demand to increase the harvest; but the wildlife in a set area does not increase, its numbers remain steady and so does the harvest if it is to be sustainable (an increase in harvest will eventually cause the extinction of the wildlife as has occurred in areas where control disintegrated with political upheaval[300]). A steady harvest means that each person now receives a declining income. Secondly, each person is not content to receive even a steady income for their whole lives, let alone a declining income—they expect an increasing income so that the demand for increasing harvest is exacerbated. Thirdly, the area set aside for CBC declines over time—I explain the reasons for this below—so that the wildlife population from which a harvest is taken decreases. For all these reasons CBC areas become unsustainable.[301] There are really only two options. One is to let the human population increase within the area, which eventually leads to unsustainable use and collapse of conservation, or the human population is limited to a set number. So far a socially acceptable mechanism to do this has been hard to find, but unless such a mechanism is found, conservation will not succeed. To some extent bushmeat hunting in tradional areas also falls into this scenario, and studies show that demand for protein results in unsustainable hunting.[302]

A second problem arises in CBC areas because part of the biota is lost, either deliberately—large predators are usually extirpated—or through over-harvesting in historical times—examples are the loss of bison in Europe and America. Species are also lost through the demands of economics to intensify agriculture. The result can be seen in the continuing loss of common bird species in the English countryside due to decline in insect food from insecticides and seeds from herbicides, and to predation from a distorted predator community.[303] This process results in loss of rare species, and greater abundance of a few common species, with an overall loss of biodiversity.[304] The consequence of the loss of species, especially loss of top predators, is a cascading instability, an unravelling of the ecosystem; this has now been demonstrated worldwide.[305]

The Ngorongoro Conservation Area, which comprises the eastern part of the Serengeti ecosystem, became the world's first CBC area in 1959 when it was partitioned from the original Serengeti Park. It was deemed a great experiment. Now fifty years later it shows all the signs of failure: originally it was set aside for Maasai pastoralists to graze their cattle. Maasai numbers have increased but cattle numbers have levelled out because there is a finite area for grazing. In addition other tribes have immigrated and taken up land. Consequently we see Maasai changing their traditional lifestyle and taking up agriculture because they are unable to obtain enough food from their cattle. Cultivation is illegal in the NCA but it is being practised anyway out of necessity. The situation is unstable and both social and political dissent is increasing for all the reasons explained previously.

The NCA adopted a policy in 1974 to protect the wildlife in the Ngorongoro Crater from the increasing human settlement there—they banned the presence of people and their livestock on the Crater floor. In effect they created a protected area to solve the problem.[306] Richard Turnbull, the last colonial governor of Tanganyika, when introducing legislation for the Ngorongoro Conservation Area in parliament, stated that if there is a conflict between people and wildlife, the interests of people will always come first. With unabated increasing human populations this policy means that in the long run there will be no wildlife and no natural resources. To paraphrase Stephen Schneider, 'We first' means 'We end'.[307]

Community-based conservation, therefore, is essential for the future conservation of world resources but it is not sufficient by itself and current practice is unsustainable. Far more innovative and daring approaches will be needed.[308]

* * *

The alternative approach to conservation is to protect areas for the primary purpose of preserving the natural biota. This is protected area conservation (PAC) and it is the traditional approach. It has the advantage of solving the thorny problem of how to prevent an increase in the number of people already living in an area by having none present in the first place. PAC is essential for the preservation of large carnivores, rare species that cannot tolerate humans; as gene banks for future uses; for providing a refuge from over-exploitation; and to act as an ecological baseline to detect unwanted changes from human activities elsewhere. Protected areas are also needed to preserve unique ecosystems such as Serengeti. In short, we can't do without them.

Nevertheless, there are several intractable problems even in protected areas that threaten the future of conservation. First, as mentioned before, there is simply not enough area protected in the world to include all species; only about half can be included in the present set of protected areas, and with burgeoning human populations this area is unlikely to increase. Individual protected areas are also not large enough to protect populations that can exist by themselves if all individuals are eliminated outside the boundaries. The result is that species will be lost from the protected area over time; there is evidence that this loss is already occurring in African protected areas.[309] Because of these limitations protected areas depend on support from outside their boundaries to maintain species diversity. I have already mentioned that ecosystem processes, such as the flow of water, are also dependent on what happens outside.

Secondly, the practice of PAC rests on delimiting an area with a legal boundary, which is then fixed for eternity. Current ecological research around the world, including our work in Serengeti, shows that ecosystems are changing as climate changes; and they have been changing for millions of years. We know this has been the case for the past 4 million years in Serengeti and such change will continue into the future (Chapter 4). So static boundaries will not accommodate the changes in distribution of plants and animals as climate changes their environment; indeed it is likely that in fifty years from now none of the current areas will be protecting what they were intended to protect— the communities will have changed their locations. The only way we can deal with this shift is to have either very large areas or a patchwork of small areas that can capture a moving ecosystem. Perhaps the British system of Sites of Scientific Interest that cover the country might be a model for such an approach but only if the patches are large enough. Unless conservation takes up this challenge protected areas will not be meeting their original goals in the long run.

A third problem with the protected area approach lies in the legal insecurity of the boundaries. Despite legal demarcation many protected areas suffer attrition of the area as demands for land increase. Boundaries are realigned to suit political whims and over time the area diminishes; we have seen this realignment of boundaries and reduction of protected area in the Mara Reserve and Maswa Reserve of the Serengeti ecosystem, and it is common practice around the world.[310] If protected areas are to serve their function then the only solution is to replace lost areas with other areas of similar biological value, a policy that is so far unrecognized and unpractised.

So we must conclude that neither CBC nor PAC is sufficient alone and both approaches are required to form a coordinated policy for sustainable conservation of world resources. Also both require new and daring policies to make them viable.

* * *

There are many reasons why mankind endeavours to conserve its precious resources but suffice it to say that most peoples recognize the need for conservation.[311] Conservation entails the maintenance of renewable natural resources, including complete biotas (the suite of living species), for the indefinite future. It does not prohibit the use of such resources, merely asks that they are used sustainably, which means they are not degraded by a loss of the original species due to such use. It does not mean that all resources everywhere must be conserved, just that a sample large enough to be viable should be maintained.

Conservation also includes the preservation of what the world deems areas, structures, artefacts, and items of great intrinsic value or rarity; these would include both natural and cultural items of great value. Cultural artefacts would include, for example, the Pyramids and Sphinx of Egypt, ancient buildings, statues, and other art objects. Natural artefacts would include fossils of great rarity, such as those found at Olduvai and Laetoli in Serengeti, and natural areas would include places of great scenic beauty such as the Grand Canyon and unique collections of natural diversity such as the Galapagos Islands. The world has already identified many such areas and has set them aside as World Heritage Sites under the care of the United Nations, namely UNESCO. Such sites have the agreement of the country in which they reside to preserve them and protect them from decay and destruction. The country has taken on the obligation on behalf of the world. The Serengeti National Park was one of the first sites in the world so identified at the United Nations Conference on the Human Environment in Stockholm in 1972; it was formally designated in 1981. The important point is that Tanzania has taken on the obligation of preserving Serengeti in as undisturbed a state as possible because of its unique character. It is recognized as unique and of great value by the world as a whole.

Do sovereign countries have the right to dispose of resources within their boundaries without regard to the wishes of the world? Humans have claimed ownership of their resources and the right to do as they wish within their national boundaries. Yet no country owns the land on which it resides in the sense of being a permanent resident; all states are temporary renters with relatively short lifetimes. The longest lasting empires such as that of Rome are now long gone, replaced by other peoples and countries. The greatest empire the world has seen, that of the Mongols, lasted a mere 300 years. The British Empire, perhaps the most globally widespread, lasted an even shorter time. Other peoples arrive to replace those who came before. Do we have the right to eradicate natural resources, extirpate species, and leave little or nothing for those to come? Sovereign countries may have the physical and self-proclaimed self-serving legal ability to destroy the unique and rare assets of the world and leave a wasteland for those to come, but they most certainly do not have the moral right. The great irony is that the future of civilization depends on humans taking second place to nature for some of the time.

Conservation is fundamentally about morality.[312] President Julius Nyerere of Tanzania understood that when he made his famous statement in 1961. Will those who follow him also understand that Tanzania holds Serengeti in trust for the rest of the world? The lesson of Serengeti is that nothing is ever secure against human greed. We must be forever vigilant. If we cannot protect Serengeti we are unlikely to protect anything else in the natural world; this is the test of conservation.

In particular, the future of Serengeti depends on funds being found for construction of the south road. Unless this occurs the Serengeti will be cut in half by a northern trunk road. It is up to the international community to find these funds.[313] Finally, it was the accumulated data from the long-term monitoring that allowed scientists to warn of the consequences of blocking the migration, and supporting those data was the crucial information from history. Without those data, and knowledge of the past, political opinion would have had free rein. However, the monitoring is coming to an end. A Serengeti Monitoring Trust Fund is now needed to continue data collection along the lines of the Galapagos Trust.[314]

FURTHER READING

There are numerous scientific papers, many of which are mentioned in the notes. Some of the main books are:

Bertram, B. C. R. 1978. *Pride of Lions*. London: Dent.

Caro, T. M. 1994. *Cheetahs of the Serengeti Plains*. Chicago: University of Chicago Press.

Frame, G. W., and Frame, L. H. 1981. *Swift and Enduring: Cheetah and Wild Dogs of the Serengeti*. New York: E. P. Dutton.

Kruuk, H. 1972. *The Spotted Hyena*. Chicago: University of Chicago Press.

Packer, C. 1994. *Into Africa*. Chicago: University of Chicago Press.

Schaller, G. B. 1972. *The Serengeti Lion*. Chicago: University of Chicago Press.

Sinclair, A. R. E. 1977. *The African Buffalo: A Study of Resource Limitation of Populations*. Chicago: University of Chicago Press.

Sinclair, A. R. E., and Norton-Griffiths, M. (eds). 1979. *Serengeti: Dynamics of an Ecosystem*. Chicago: University of Chicago Press.

Sinclair, A. R. E., and Arcese, P. (eds). 1995. *Serengeti II: Dynamics, Management and Conservation of an Ecosystem*. Chicago: University of Chicago Press.

Sinclair, A. R. E., Packer, C., Mduma, S. A. R., and Fryxell, J. M. (eds). 2008. *Serengeti III: Human Impacts on Ecosystem Dynamics*. Chicago: University of Chicago Press.

APPENDIX

The Main Species in Serengeti

COMMON SPECIES OF THE SERENGETI

Here I provide a short summary of the main herbivore species in the Serengeti eco-system. The orders Proboscidea, Hyracoidea, and Tubulidentata have recently been placed in the Superorder Afrotheria, a group that evolved within Africa. The full list with Latin names follows.

WILDEBEEST

Wildebeest numbers were around 200,000 in the late 1950s. They were presumably about half this in the 1930s. Numbers increased rapidly after rinderpest died out in 1963 and reached 1.4 million by 1977. The population levelled out at 1.3 million and has remained there since 1977 apart from a drop to 800,000 in 1993–4 during the great drought of 1993. Numbers appear to have stabilized, having recovered from the drought, and the census of 2009 showed 1.3 million in the population. They are grass feeders, preferring short grass leaves.

ZEBRA

This species has remained at approximately the same level of 200,000 since counts began in 1966. The latest census showed 184,000 in 2003. Part of the population remains scattered through the savanna and may be resident there but information is lacking. The other part migrates with the wildebeest although it stays further west on the plains around the Simba kopje to Moru kopje line. They feed entirely on course grass, often eating the dry stems in the dry season.

THOMSON'S GAZELLE

Trends suggest that numbers have declined once wildebeest reached their stable level in 1977. Numbers were around 600,000 in 1971 and 300,000 in 1996. They migrate following the wildebeest as the herds leave the plains in May–June, using areas already grazed by wildebeest. They remain in the western and central woodlands and do not move north when the wildebeest go to Kenya. They feed on a mixture of grasses and forbs.

GRANT'S GAZELLE

An earlier rough estimate of 50,000 in 1982 is similar to the count of 55,000 in 2003. Grants are common on the open plains even in the dry season for they can survive without free water, getting their moisture from their food, which is entirely forbs. They occur throughout the savanna in small numbers and there may be some movement from the plains to nearby woodlands in the dry season.

TOPI

Numbers have not changed much from 55,000 in 1971 compared with a recent count of 38,990 in 2003. They prefer the wetter areas in the corridor where herds of several thousand occur on the Ndoho, Dutwa, and Ruwana plains. In the central and northern woodlands they occur in small herds of 10 or less and often mix with the closely related kongoni. They feed on leaves of tall grasses.

KONGONI

Similarly this species has not changed in number since 1971. However, numbers are still very small, around 8,000. They prefer the drier eastern woodlands and long grass plains where they occur in herds up to fifty. They, like topi, feed on leaves of tall grasses.

IMPALA

Impala, the quintessential antelope, is, however, a species that is not closely related to any other antelope; it separated a long way back in evolutionary time. It is the commonest antelope of the savanna (about 90,000), but never ventures into grassland, especially the plains. It feeds on grass in the wet season and shrubs and herbs in the dry season. Where thickets have increased through woodland regrowth densities appear to have increased.

WATERBUCK

This species is confined to the major rivers with permanent water. Numbers are very low, probably not more than 1,500. They live in tall grassland and forest and feed on grass.

AFRICAN BUFFALO

The buffalo is a savanna species that needs to drink every day and so must be near permanent water. In Serengeti they feed on tall grasses. The population of this species, like that of wildebeest, recovered after the rinderpest disappeared in 1963, but then collapsed through both poaching and drought in the period 1978–94. They are recovering slowly in the southern half of Serengeti, but not in northern Serengeti or Mara Reserve. They reached 40,000 in 2011.

GIRAFFE

Giraffe feed on low shrubs but can use tall trees if necessary. Numbers are around 10,000. Trends are not clear. There is an indication that poaching in the first decade of the 2000s has targeted this species and numbers could be declining.

ELEPHANT

Elephant were probably present in Serengeti in the mid-nineteenth century but disappeared when ivory hunters almost exterminated the species in East Africa during 1840–90. Elephants were present only in the Masirori swamp on the Mara River and none were known in the rest of the area in 1913. Two were seen by Waikoma elders in 1930. Two populations invaded the ecosystem in the 1950s, one from the north in Kenya and the other from the south in Maswa. The two populations met in the centre of the park by the late 1960s. Numbers reached 3,000 by 1977 but collapsed in the second period of high ivory poaching (1973–88), reaching a low point of 400 animals in 1986. Since 1990 they have recovered very fast and by 2011 reached 3,000.

ELAND

This is one of the migrant species that depend on the north-western Serengeti in the dry season where there are sufficient bushes for their food. In the wet season they migrate to the plains and feed on grasses. There are some 15,000 animals.

ROAN ANTELOPE

A small number of about fifty used to occur in north-western Serengeti and Mara but numbers have declined probably to less than 10. In 1913 they were found right across from the Ikorongo Hills in the west to near Waso in the east. Small herds were also present around Banagi and Nyaraswiga Hill in the period 1930–80 but the last one was seen in 1981. A small herd survived around Mangwesi Hill and this has increased a small amount with the protection afforded by the Grumeti Game Reserve in the 2000s. They are restricted to *Combretum* and *Terminalia* woodland, which is found on the hills and north-western Serengeti ridge tops. They feed on the *Hyparrhenia–Loudetia* grasses. Roan is relatively more abundant in the far south at Maswa around kopjes.

ORIBI

Oribi occur mainly in the *Terminalia* woodland with over 6-foot-high *Hyparrhenia* grass in north-west Serengeti, with a few on rocky ridges in the north-east. They number around 7,000. They are the most abundant of the small antelopes in the appropriate habitat.

KLIPSPRINGER

A rock-inhabiting species on kopjes of the woodlands but not the plains, they occur singly or in small family groups.

STEINBUCK

These occur in singles or pairs at very low density in the dry eastern woodlands and around the edge of the plains.

GREY DUIKER

Duiker occur in broad-leaved woodland, both in *Combretum* on rocky hills and in the *Terminalia* woodland. They are always in singles and at very low density.

BLACK RHINOCEROS

Rhino were about 500 in number before 1977. They were almost exterminated by 1978. Currently there are 30 in the south and 10 in the north of the system.

ORYX

These occur only on the semi-arid Salai plains of the NCA and Loliondo. Occasionally they wander into eastern Serengeti National Park. Their numbers are very low.

REEDBUCK

Two species occur. Bohor occur commonly throughout the long grass plains, as recorded in night road transects, but they are rarely seen by day. They are also seen along rivers in daylight hours throughout the woodlands. Mountain (Chandler's) reedbuck occur in small groups or singly on top of the highest hills—Kuka, Lobo, Nyaraswiga, Nyaroboro, and Itonjo.

KUDU

Both species occur in very small numbers at the extremes of the ecosystem. Greater kudu occur in the kopjes at the southern end of Maswa. Lesser kudu occur in the mountain thickets and forests of the Loita Hills, east of the Mara Reserve in Kenya.

PIGS

Warthog are ubiquitous though in low numbers on the eastern plains. They appear to have dropped in number in recent years. Bushpig occur in the riverine forests but are rarely seen. Myles Turner observed one Giant Forest Hog in 1972 in the riverine forest of the Mbalipali, 4.5 miles west of Kogatende. He assumed it had wandered downstream along the Mara from the Loita forests.

HYRAX

Three species occur. Bush hyrax (*Heterohyrax*) feed on dicots and climb trees, whereas rock hyrax (*Procavia*) feed on monocots and do not climb trees. Both are widespread on most but not all woodland kopjes and those at the edge of the plains. Kopjes further into the plains are without hyrax. Tree hyrax (*Dendrohyrax*) live in the montane forests of the Mara watershed but not in the lowland forests of the Grumeti and Mbalageti.

HARES

Three species occur. Cape hares occur on the Serengeti plains, whereas Crawshay's hare lives in the woodlands. They are indistinguishable in the field. The red rock hare is very localized, recorded so far only on top of Kuka Hill, but common where found.

PRIMATES

Olive baboon and vervet monkeys are ubiquitous in the woodlands. They are most abundant associated with the Grumeti and Mbalageti Rivers. Black-and-white colobus are also restricted to the forests of those rivers. Greater galago are restricted to riverine forest of both the north and west. Bushbabies occur throughout the woodlands.

LIST OF MAMMALS

Common Name	Latin Name

ORDER INSECTIVORA

Four-toed hedgehog — *Erinaceus albiventris*

ORDER PRIMATES

Black-and-white colobus	*Colobus guereza*
Olive baboon	*Papio cynocephalus*
Patas monkey, Ikoma	*Erythrocebus patas baumstarki*
Vervet monkey	*Cercopithecus aethiops*
Greater galago	*Galago crassicaudatus*
Bushbaby, lesser	*Galago senegalensis*

ORDER CARNIVORA

Lion	*Panthera leo*
Spotted hyena	*Crocuta crocuta*
Cheetah	*Acinonyx jubatus*
Leopard	*Panthera pardus*
Wild dog, African	*Lycaon pictus*
Black-backed jackal	*Canis mesomelas*
Golden jackal	*Canis aureus*
Side-striped jackal	*Canis adustus*
Bat-eared fox	*Otocyon megalotis*
Striped hyena	*Hyaena hyaena*
Aardwolf	*Proteles cristatus*
Wildcat	*Felis silvestris*
Serval	*Leptailurus serval*
Caracal	*Caracal caracal*
Egyptian (Grt grey) mongoose	*Herpestes ichneumon*

Banded mongoose	*Mungos mungo*
Dwarf mongoose	*Helogale undulata*
Black-tipped mongoose	*Herpestes sanguineus*
Genet, Common	*Genetta genetta*
Genet, Spotted	*Genetta tigrina*
White-tailed mongoose	*Ichneumia albicaudata*
Marsh mongoose	*Herpestes paludinosus*
Civet, African	*Viverra civetta*
Ratel, Honey badger	*Mellivora capensis*
Zorilla	*Ictonyx striata*
Palm civet	*Nandinia binotata*
Spotted-necked otter	*Lutra masculicollis*
Cape clawless otter	*Aonyx capensis*

ORDER PROBOSCIDEA

Elephant	*Loxodonta africana*

ORDER HYRACOIDEA

Bush hyrax	*Heterohyrax brucei*
Rock hyrax	*Procavia capensis*
Tree hyrax	*Dendrohyrax arboreus*

ORDER TUBULIDENTATA

Aardvark	*Orycteropus afer*

ORDER PERISSODACTYLA

Zebra, Burchell's	*Equus burchelli*
Rhinoceros, Black	*Diceros bicornis*

ORDER ARTIODACTYLA

Giraffe	*Giraffa camelopardalus*
Buffalo, African	*Syncerus caffer*
Eland	*Taurotragus oryx*
Topi	*Damaliscus lunatus*
Kongoni (Coke's hartebeest)	*Alcelaphus buselaphus*
Wildebeest	*Connochaetes taurinus*
Impala	*Aepyceros melampus*
Defassa waterbuck	*Kobus defassa*

(continued)

(Continued)

Common Name	Latin Name
Bohor reedbuck	*Redunca redunca*
Mountain reedbuck	*Redunca fulvorufula*
Thomson's gazelle	*Gazella thomsoni*
Grant's gazelle	*Gazella granti*
Oryx (Fringe-eared)	*Oryx beisa*
Roan antelope	*Hippotragus equinus*
Oribi	*Ourebia ourebi*
Klipspringer	*Oreotragus oreotragus*
Dikdik (Kirk's)	*Madoqua kirki*
Steinbuck	*Raphicerus campestris*
Grey duiker	*Sylvicapra grimmia*
Hippopotamus	*Hippopotamus amphibius*
Greater kudu	*Tragelaphus strepsiceros*
Lesser kudu	*Tragelaphus imberbis*
Bushbuck	*Tragelaphus scriptus*
Warthog	*Phacochoerus aethiopicus*
Bushpig	*Potamochoerus porcus*
Giant forest hog	*Hylochoerus meinertzhageni*

ORDER PHOLIDOTA

Pangolin, Ground	*Manis temmicki*

ORDER RODENTIA

Spring hare	*Pedetes capensis*
Cape crested porcupine	*Hystrix africae-australis*
North African crested porcupine	*Hystrix cristata*

ORDER LAGOMORPHA

Cape hare	*Lepus capensis*
Crawshay's hare	*Lepus crawshayi*
Red rock hare	*Pronolagus rupestris*

NOTES

Chapter 1: Serengeti: A Wonder of the Natural World

1. In 2011 the Latin name *Acacia* was transferred from the African type genus to one in Australia. All of the African species have new generic names. However, the name acacia is so quintessentially African that I will continue to use it as a common name by calling them all African acacias.

2. Murray Watson conducted his PhD research on the migration and population dynamics of Serengeti wildebeest at the University of Cambridge, 1962–6. He later became an ecological consultant, specializing in aerial census of wildlife.

Chapter 2: The Great Migration

3. For a review on the status of global migration systems see Harris, G., Thirgood, S., Hopcraft, J. G. C., Cromsigt, J. P. G. M., and Berger, J. (2009), 'Global Decline in Aggregated Migrations of Large Terrestrial Mammals', *Endangered Species Research* 7: 55–76.

4. H. M. Stanley shows in his book *In Darkest Africa* numerous maps from 100 BC to AD 1819 that illustrate the course of the Nile and two lakes as sources somewhere in the centre of Africa. Although these cartographers probably copied from each other it is also possible that travellers had reported that the Nile originated from two great lakes. The Al Adrisi map from 1160 shows Lake Victoria, its tributaries, and the source of the Nile. See Stanley, H. M. 1890, *In Darkest Africa* (London: Sampson, Low, Marston, Searle and Rivington).

5. For information on the riverine forests see Sharam, G., Sinclair, A. R. E., and Turkington, R. (2006), 'Establishment of Broad-Leaved Thickets in Serengeti, Tanzania: The Influence of Fire, Browsers, Grass Competition and Elephants', *Biotropica* 38: 599–605; Sharam, G., Sinclair, A. R. E., Turkington, R., and Jacob, A. L. (2009), 'The Savanna Tree *Acacia polyacantha* Facilitates the Establishment of Riparian Forests in Serengeti National Park, Tanzania', *Journal of Tropical Ecology* 25: 31–40; Sharam, G., Sinclair, A. R. E., and Turkington, R. (2009), 'Serengeti Birds Maintain Forests by Inhibiting Seed Predators', *Science* 325: 51.

6. On what drives the migration see: Fryxell, J. M. (1995), 'Aggregation and Migration by Grazing Ungulates in Relation to Resources and Predators', and Murray, M. G. (1995),

'Specific Nutrient Requirements and Migration of Wildebeest', in A. R. E. Sinclair and P. Arcese (eds), *Serengeti II: Dynamics, Management and Conservation of an Ecosystem*, pp. 257–73, 231–56 (Chicago: University of Chicago Press); Fryxell, J. M., Wilmshurst, J. F., Sinclair, A. R. E., Haydon, D. T., Holt, R. D., and Abrams, P. A. (2005), 'Landscape Scale, Heterogeneity, and the Viability of Serengeti Grazers', *Ecology Letters* 8: 328–335; Fryxell, J. M., Wilmshurst, J. F., and Sinclair, A. R. E. (2004), 'Predictive Models of Movement by Serengeti Grazers', *Ecology* 85: 2429–35; Holdo, R. M., Holt, R. D., and Fryxell, J. M. (2009), 'Opposing Rainfall and Nutrient Gradients Best Explain the Wildebeest Migration in the Serengeti', *American Naturalist* 173: 431–45.

7. For facilitation by wildebeest on gazelle see McNaughton, S. J. (1976), 'Serengeti Migratory Wildebeest: Facilitation of Energy Flow by Grazing', *Science* 191: 92–4.

8. See Sinclair, A. R. E. (1977), 'The Lunar Cycle and the Timing of Conception in Serengeti Wildebeest', *Nature* 267: 832–3.

9. For wildebeest birth adaptations see pp. 103–6 in: Sinclair, A. R. E. (1977), *The African Buffalo: A Study of Resource Limitation of Populations* (Chicago: University of Chicago Press).

10. Richard Estes recounts calves finding their mothers in Estes, R. D. (1991), *The Behavior Guide to African Mammals: Including Hoofed Mammals, Carnivores, Primates* (Berkeley: University of California Press).

11. For impala behaviour see Jarman, P. J., and Jarman, M. V. (1973), 'Social Behaviour, Population Structure, and Reproductive Potential in Impala', *East African Wildlife Journal* 11: 329–38.

12. For papers on oribi in Serengeti see Arcese, P., Jongejan G., and Sinclair A. R. E. (1995), 'Behavioural Flexibility in Small African Antelope: The Size and Composition of Oribi Groups', *Ethology* 99: 1–23; Arcese, P. (1994), 'Harem Size and Horn Symetry in Oribi', *Animal Behavior* 48: 1485–8; Mduma, S. A. R. (1995), 'Distribution and Abundance of Oribi, a Small Antelope', in Sinclair and Arcese (eds), *Serengeti II*, pp. 220–30.

13. For buffalo habitat choice and distribution see pp. 56–60 and 202 in Sinclair, *The African Buffalo*.

14. Publications on lions in Serengeti are numerous. See especially Schaller, G. B. (1972), *The Serengeti Lion* (Chicago: University of Chicago Press); Bertram, B. C. R. (1973), 'Lion Population Regulation', *East African Wildlife Journal* 11: 215–25; Hanby, J. P., Bygott, J. D., and Packer, C. (1995), 'Ecology, Demography, and Behaviour of Lions in Two Contrasting Habitats: Ngorongoro Crater and Serengeti Plains', and Scheel, D., and Packer, C. (1995), 'Variation in Predation by Lions: Tracking a Movable Feast', in Sinclair and Arcese (eds), *Serengeti II*, pp. 315–31, 299–314; Hopcraft, J. G. C., Sinclair, A. R. E., and Packer C. (2005), 'Prey Accessibility Outweighs Prey Abundance for the Location of Hunts in Serengeti Lions', *Journal of Animal Ecology* 74: 559–66; Packer, C., Hilborn, R., Mosser, A., Kissui, B., Borner, M., Hopcraft, G., Wilmshurst, J., Mduma, S., and Sinclair, A. R. E. (2005),

'Ecological Change, Group Territoriality and Population Dynamics in Serengeti Lions', *Science* 307: 390–3.

15. Important publications on hyenas are Kruuk, H. (1972), *The Spotted Hyena* (Chicago: University of Chicago Press); Frank, L. G., Holekamp, K. E., and Smale, L. (1995), 'Dominance, Demography, and Reproductive Success of Female Spotted Hyena', and Hofer, H., and East, M. (1995), 'Population Dynamics, Population Size, and the Commuting System of Serengeti Spotted Hyena', in Sinclair and Arcese (eds), *Serengeti II*, pp. 364–84, 332–363.

Chapter 3: African Buffalo

16. Early papers proposed that weather alone limited animal population numbers, or a variety of factors that depended on weather. These included Uvarov, B. P. (1931), 'Insects and Climate', *Transactions of the Entomological Society of London* 79: 1–249; and Thompson, W. R. (1939), 'Biological Control and the Theories of the Interactions of Populations', *Parasitology* 31: 299–388. Andrewartha and Birch developed the theory of environmental limiting factors in Australia where both weather and populations fluctuate markedly. They were concerned with explaining these fluctuations. See Andrewartha, H. G., and Birch, L. C. (1954), *The Distribution and Abundance of Animals* (Chicago: University of Chicago Press).

17. A. J. Nicholson revolutionized thinking about population limitation when he introduced the concept of regulation through density-dependent factors. This idea was stimulated by his observations also on Australian insects that rarely went extinct. This suggested that some form of regulation mechanism was acting so that when numbers went down the regulator was relaxed and when numbers rose, it was imposed more severely. This would result in numbers being in a 'state of balance with the environment'. Factors that could act in this way are competition between animals for food or space, or predation. However, there was little evidence in nature on these processes and those who championed the weather ideas discounted the regulation theory. By 1958 there was much debate but little evidence one way or the other. See Nicholson, A. J. (1933), 'The Balance of Animal Populations', *Journal of Animal Ecology* 2: 132–78. I detail the history of the regulation theory in Sinclair, A. R. E. (1989), 'Population Regulation in Animals', in J. M. Cherrett (ed.), *Ecological Concepts* pp. 197–241 (Oxford: Blackwell Scientific).

18. Field, C. R. (1968), 'The Food Habits of Some Wild Ungulates in Uganda', PhD thesis, Cambridge University. Field, C. R. (1968), 'A Comparative Study of the Food Habits of Some Wild Ungulates in Queen Elizabeth Park, Uganda: Preliminary Report', in M. A. Crawford (ed.), *Comparative Nutrition of Wild Animals*, pp. 135–51. Symposium of the Zoological Society of London, no. 21 (London: Academic Press); Grimsdell, J. J. R. (1969), 'The Ecology of the Buffalo, *Syncerus caffer*, in Western Uganda', PhD thesis, Cambridge University.

19. Murray Watson was one of the three scientists constituting the Serengeti Research Program (1962–6) funded by the Food and Agriculture Organization of the United Nations. He made a career in aerial survey in various countries of Africa and Asia. See Watson, R. M. (1967), 'The Population Ecology of the Serengeti Wildebeest', PhD thesis, Cambridge University.

20. Patrick Duncan studied the ecology of topi in western Serengeti in the early 1970s. He moved to France to make his career as a government ecologist and eventually became a senior scientist there. See Duncan, P. (1975), 'Topi and Their Food Supply', PhD dissertation, University of Nairobi, Nairobi, Kenya.

21. See Schaller, George B. (1964, 1988), *The Year of the Gorilla* (Chicago: University of Chicago Press); (1967), *The Deer and the Tiger* (Chicago: University of Chicago Press).

22. Hans Kruuk conducted the first studies on behaviour and ecology of spotted hyenas in the Serengeti and Ngorongoro from 1964 to 1971. He was the Deputy Director of the Serengeti Research Institute 1968–71. After two years back at Oxford with Niko Tinbergen he moved to the British Government's Institute of Terrestrial Ecology in the Scottish Highlands where he remained until retirement in 1997. He was also a Professor at Aberdeen University. He became a world expert on the ecology of Eurasian badgers and otters. Other research included platypus in Australia, feral dogs in Galapagos, and various species of otters and badgers in Asia, Africa, and South America from which he has written many books. See Kruuk, H. (1972), *The Spotted Hyena* (Chicago: University of Chicago Press).

23. Niko Tinbergen, Professor in Animal Behaviour at Oxford University, won the Nobel Prize for his studies in animal behaviour in 1973, shared with Konrad Lorenz and Karl von Frisch.

24. A couple of decades after this incident Hans Kruuk related the story to Michael Crichton, who incorporated it into his novel.

25. George Schaller was the first to study lions in the Serengeti in the 1960s. He went on to become a senior scientist with the New York Zoological Society, later to become the Wildlife Conservation Society, New York. His main interests then focused on Asia, especially Mongolia, Laos, Tajikistan, Afghanistan, and China, but he has also worked in Brazil, documenting the wildlife in a long series of books. He was instrumental in setting up a major protected area in Tibet, the Chang Tang Reserve on the Tibetan Plateau at about 1.16 million square miles, and he continues to study the wildlife there. He won the prestigious Cosmos Prize in Japan and since 2008 has been vice-president of Panthera, a conservation group focusing on carnivores. He has published a number of books with the University of Chicago Press. See Schaller, G. B. (1972), *The Serengeti Lion*; also (1993), *The Last Panda*; with Hu Jinchu, Pan Wenski, and Zhu Jing (1985), *The Giant Pandas of Wolong*; and (1998) *Wildlife of the Tibetan Steppe*.

26. Hugh Lamprey was the Game Warden for Tanganyika in the early 1950s. During this time he documented for the first time the habitat partitioning of ungulates in East Africa, focusing his studies on Tarangire National Park. After a period as Principal of Mweka Wildlife College, Moshi, he was appointed the first Director of the Serengeti Research Institute, 1966–72. He went on to study the ecology of pastoralists in Northern Kenya until he retired to the UK in 1987. See Lamprey, H. F. (1963), 'Ecological Separation of the Large Mammals in the Tarangire Game Reserve, Tanganyika', *East African Wildlife Journal* 1: 63–92.

27. Peter and Mattie Jarman initially studied the social organization of ungulates at Lake Kariba in the then Southern Rhodesia (Zimbabwe) and from which Peter produced the classic work on behavioural ecology and sociobiology (Jarman, P. J. (1974). 'The Social Organization of Antelope in Relation to Their Ecology', *Behaviour* 48: 215–66; also Jarman, P. J. (1972), 'Seasonal Distribution of the Large-Mammal Populations of the Unflooded Middle Zambezi Valley', *Journal of Applied Ecology* 9: 283–99). They came to Serengeti to study impala in 1968–71. See Jarman, P. J., and Jarman, M. V. (1973), 'Social Behavior, Population Structure, and Reproductive Potential in Impala', *East African Wildlife Journal* 11: 329–38. They emigrated to Australia in 1974 where Peter became a lecturer at the University of New England, Armidale, NSW.

28. I describe in greater detail the concept of the grazing succession, Vesey-Fitzgerald's theory, and Richard Bell's research in Chapter 9.

29. Sinclair, A. R. E., and Gwynne, M. D. (1972), 'Food Selection and Competition in the East African Buffalo (*Syncerus caffer* Sparrman)', *East African Wildlife Journal* 10: 77–89.

30. The main results of the buffalo research are in Sinclair, *The African Buffalo*.

Chapter 4: The Great Pandemic

31. Peters, C. R., Blumenschine, R. J., Hay, R. L., Livingstone, D. A., Marean, C. W., Harrison, T., Armour-Chelu, M., Andrews, P., Bernor, R. L., Bonnefille, R., and Werdelin, L. (2008), 'Paleoecology of the Serengeti-Mara Ecosystem', in A. R. E. Sinclair, C. Packer, S. A. R. Mduma, and J. M. Fryxell (eds), *Serengeti III: Human Impacts on Ecosystem Dynamics*, pp. 47–94 (Chicago: University of Chicago Press); Andrews, P., and Bamford, M. (2008), 'Past and Present Vegetation Ecology of Laetoli, Tanzania', *Journal of Human Evolution* 54: 78–98.

32. Morell, V. 1995. *Ancestral Passions: The Leakey Family and the Quest for Humankind's Beginnings* (New York: Simon and Schuster).

33. Peters et al., 'Paleoecology of the Serengeti-Mara Ecosystem'; Andrews and Bamford, 'Past and Present Vegetation Ecology of Laetoli, Tanzania'.

34. Gentry, A. W. (1967), '*Pelorovis olduwayensis* Reck, an Extinct Bovid from East Africa', *Bulletin of the British Museum (Natural History), Geology* 14: 245–99.

35. Peters et al., 'Paleoecology of the Serengeti-Mara Ecosystem'; Andrews and Bamford, 'Past and Present Vegetation Ecology of Laetoli, Tanzania'.

36. Leakey, M. D., and Hay, R. L. (1979), 'Pliocene Footprints in the Laetoli Beds at Laetoli, Northern Tanzania', *Nature* 278: 317–23.

37. From conversations with Raymonde Bonnefille in 2001. See also Bonnefille, R., Hamilton, A. C., Linder, H. P., and Riolet, G. (1990), '30,000-Year-Old Fossil Restionaceae Pollen from Central Africa and its Biogeographical Significance', *Journal of Biogeography* 17: 307–14; Bonnefille, R., and Chalie, F. (2000), 'Pollen Inferred Precipitation Time-Series from Equatorial Mountains, Africa, the Last 40 kyr BP', *Global and Planetary Change* 26: 25–50.

38. Oliver, R., and Mathew, G. (1963), *History of East Africa*, vol. 1 (Oxford: Clarendon Press).

39. Davidson, B. (1992), *Africa in History: Themes and Outlines* (London: Phoenix Press).

40. Accounts of the movements of the lake Nilotics come from interviews with elders of the Luo tribe in Tanzania by Elijah Amworo and given to me.

41. History of the Wasukuma is documented in the Bujora Sukuma Catholic Mission Museum, Kisesa, outside Mwanza.

42. S. E. White stated in 1913 that tribes were limited by tsetse fly west of the Ikorongo hills. White, S. E. (1915), *The Rediscovered Country* (New York: Doubleday, Page and Co.)

43. Recounted in Turner, Myles (1987), *My Serengeti Years*, ed. Brian Jackman (London: Elm Tree Books).

44. This account is from Stephen Makacha, an elder of the Waikoma tribe, who lives in Robanda. Stephen worked for me on and off from 1967 until 2007.

45. Wakefield, T. (1870), 'Routes of Native Caravans from the Coast to the Interior of East Africa', *Journal of the Royal Geographical Society* 11: 303–38. Wakefield, T. (1882), 'New Routes through Masai Country', *Proceedings of the Royal Geographical Society*, new series 4: 742–7. Farler, J. P. (1882), 'Native Routes in East Africa from Pangani to the Masai Country and the Victoria Nyanza', *Proceedings of the Royal Geographical Society*, new series 4: 730–42.

46. The Maasai displaced what they called the Il Datwa (spelling is variable) from the Crater Highlands sometime after 1850. Also many of the wells on the Serengeti plains are attributed by the Maasai to the Il Datwa. H. St. J. Grant considered these peoples to be the Irawq, now in Mbulu. Waller, in Homewood and Rodgers, considered them to be the earlier pastoralist Barabaig. From notes of H. St. J. Grant, District Officer at Loliondo in the 1950s. 'Masai History and Mode of Life: A Summary Prepared for the Committee of Enquiry into the Serengeti National Park' (1957), 21 pp typed. Homewood, K. M., and Rodgers, W. A. (1991), *Maasailand Ecology* (Cambridge: Cambridge University Press). See also n. 120 (Chapter 6).

47. Organ, G. E., and Fosbrooke, H. A. (1963), *Ngorongoro's First Visitor* (Dar es Salaam, Tanzania: East African Literature Bureau).

48. Wakefield, 'Routes of Native Caravans'; Wakefield, 'New Routes through Masai Country', Farler, 'Native Routes in East Africa'. Fosbrooke shows the boundary

of the Maasai at Simba Kopjes (taken from Farler, 'Native Routes in East Africa'). See Fosbrooke, H. A. (1968), 'Elephants in the Serengeti National Park: An Early Record', *East African Wildlife Journal* 6: 150–2.

49. Gray, Sir John (1957), 'Trading Expeditions from the Coast to Lakes Tanganyika and Victoria before 1857', *Tanganyika Notes and Records* 49: 226–46.

50. Turner, *My Serengeti Years*; Grant, 'Masai History and Mode of Life'.

51. Baumann, O. (1894), *Durch Massailand zur Nilquelle* (Berlin; repr. 1968, New York: Johnson Reprint Corp.).

52. He camped at Lake Lagarja, where there were 'genuine examples of nyika trees' which probably refers to the umbrella acacias (*Acacia tortilis*), classic examples of savanna trees that surround the lake today. On 29 March he left the lake, heading north-west across the plains, dry, dusty, and treeless, towards a mountain he called 'Kiruwassile', high flat-topped and with a line of small rocky hills in front of it, where he arrived the next day. This is clearly the escarpment of Nyaraboro Hill, the highest in the ecosystem, with Moru kopjes in front of it where he camped. He left the Maasai behind east of Lake Lagarja, and no further mention of them is made. Moru kopjes are described as covered in Euphorbias. These are members of the Euphorbiacae, large multibranching trees common on kopjes at Moru, and in the west of the Serengeti. The kopjes had dense, lush thornbush, much as today. In the following three days he travelled north through 'parkland', which implies large trees, with little or no regeneration scattered through grassland.

53. He was confused as to which river drainage he was on. He thought he was on the Simiyu drainage, which is south of Moru, so this could not be correct. He must have been on the Mbalageti drainage, passing Lake Magadi, which was probably dry, and camped at the swamp of Loyangalani north of Lake Magadi. He had two days' march from there to the Orangi River where there was 'gallery forest'. Since this riverine forest does not occur upstream of Kimerishi Hill, it is likely his route from Moru had been that of the earlier ivory traders in the 1870s, passing west of Mukoma Hill. He describes his descent through a lush, grassy green land, probably through the valley leading to the Mareo River, and thence to Kimerishi Hill. He crossed open plains with dry gullies lined with thornbush, which we call the Mareo plains.

54. The name Nyasirori still exists today but is applied to a guard post on the Sibora plains. The original village ('Nyasiro') was on a hill overlooking the Grumeti River in the present Grumeti Game Reserve some miles east. This is either the present Fort Ikoma village or a previous one on a hill close by downstream.

55. Baumann, *Durch Massailand zur Nilquelle*, p. 140.

56. Joseph Osgood of Salem, Massachusetts, visited Zanzibar in 1846 and stated that the 'increasing demand for ivory...favor the opinion that before long this valuable and noble animal will be exterminated in this part of Africa'. And 'ivory, borne upon the backs of slaves' in thousands. Reported in Gray, 'Trading Expeditions', 226–46.

57. Sulivan, G. L. (1873, repr. 2003), *Dhow Chasing in Zanzibar Waters* (Zanzibar: The Gallery Publications).

58. Wakefield, 'Routes of Native Caravans'; Wakefield, 'New Routes through Masai Country'; Farler, 'Native Routes in East Africa'.

59. The name Kiribasili on Farler's 1882 map is similar to that on Baumann's 1891 map. The name is no longer in use but is clearly Nyaraboro Hill.

60. Farler, 'Native Routes in East Africa', p. 736.

61. Sheriff, A. (1987), *Slaves, Spices and Ivory in Zanzibar* (Oxford: James Currey).

62. Information on the Arab traders, ivory hunting, and the slave trade comes from the first-hand account of Tippu Tip, the pre-eminent trader in Zanzibar during the second half of the nineteenth century. His memoirs were written down by Brode just before Tippu Tip died. See Brode, H. (1903; repr. 2000), *Tippu Tip: The Story of His Career in Zanzibar and Central Africa* (Zanzibar: The Gallery Publications); Spinage, C. A. (1973), 'A Review of Ivory Exploitation and Elephant Population Trends in Africa', *East African Wildlife Journal* 11: 281–9.

63. From Kjekshus, H. (1977), *Ecology Control and Economic Development in East Africa* (London: Heinemann Educational Books), as reported in Coulson, A. (1982), *Tanzania, a Political Economy* (Oxford: Clarendon Press), p. 25.

64. Sheriff, *Slaves, Spices and Ivory*.

65. Patterson, J. H. (1907; repr. 1996), *The Maneaters of Tsavo* (New York: Pocket Books).

66. White, *The Rediscovered Country*.

67. Eastman, G. (1927), *Chronicles of an African Trip* (Rochester, NY: privately printed for the author by J. P. Smith, Co.); Johnson, M. (1929), *Lion: African Adventure with the King of Beasts* (New York: G. P. Putnam's Sons); White, *The Rediscovered Country*.

68. Stigand, C. H. (1913), *Hunting the Elephant in Africa* (London: MacMillan); Bell, W. M. D. (1949), *Karamojo Safari* (New York: Harcourt, Brace).

69. Milner-Gulland, E. J., and Mace, R. H. (1991), 'The Impact of the Ivory Trade on the African Elephant Population, as Assessed by Data from the Trade', *Biological Conservation* 55: 215–29. Milner-Gulland, E. J., and Beddington, J. R. (1993), 'The Exploitation of Elephants for the Ivory Trade: An Historical Perspective', *Proceedings of the Royal Society, London, B* 252: 29–37.

70. Herne, B. (1999), *White Hunters: The Golden Age of African Safaris* (New York: Henry Holt and Co.), p. 27.

71. Justin Hando, Chief Park Warden, Serengeti, 1998–2006. Reports in 2006 from conversations with elders in the Bunda district, western Serengeti.

72. Grzimek, M., and Grzimek, B. (1960), 'A Study of the Game of the Serengeti Plains', *Zeitschrift fur Saugetierkunde* 25: 1–61; Stewart, D. R. M. (1962), 'Census of Wildlife on the Serengeti, Mara and Loita Plains', *East African Agricultural and Forestry Journal* 28: 58–60; Lamprey, H. F., Turner, M. I. M., and Bell, R. H. V. (1967), 'Invasion of the Serengeti National Park by Elephants', *East African Wildlife Journal* 5: 151–66.

73. Milner-Gulland and Mace, 'The Impact of the Ivory Trade', 215–29. Milner-Gulland and Beddington, 'The Exploitation of Elephants for the Ivory Trade', 29–37.

74. Pankhurst, R. (1966), 'The Great Ethiopian Famine of 1888–1892: A New Assessment. Part 2', *Journal of the History of Medicine* July, 271–94; Waller, R. D. (1988), 'Emutai: Crisis and Response in Maasailand 1883–1902', in D. Johnson and D. Anderson (eds), *The Ecology of Survival: Case Studies from Northeast African History*, pp. 73–114, Boulder, CO: Westview Press; Spinage, C. A. (2003), *Cattle Plague: A History* (Dordrecht: Kluwer Academic/Plenum Press).

75. Kjekshus, *Ecology Control and Economic Development*, quoted in Coulson, *Tanzania, a Political Economy*, p. 29.

76. Baumann, *Durch Massailand zur Nilquelle*, pp. 31–2. This passage is translated into English and quoted here from Grzimek, B., and Grzimek, M. (1960), *Serengeti Shall Not Die* (London: Hamish Hamilton), pp. 50, 51.

77. Smith, G. E. (1907), 'From the Victoria Nyanza to Kilimanjaro', *The Geographical Journal* 29: 249–69.

78. Ford, J. (1971). *The Role of Trypanosomiases in African Ecology* (Oxford: Clarendon Press).

79. Huxley, J. (1936), *Africa View* (London: Chatto and Windus).

80. Speke, J. H. (1863). *Journal of the Discovery of the Sources of the Nile* (London and Edinburgh: Harper and Bros.).

81. Baumann, *Durch Massailand zur Nilquelle*. He explicitly used the word 'Seuche', meaning epidemic (p. 36).

82. Quoted in Lydekker, R. (1908), *The Game Animals of Africa* (London: Rowland Ward). See also Millais, J. G. (1918; repr. 2006), *Life of Frederick Courtenay Selous* (Zanzibar: Gallery Publications).

83. Hinde, S. L., and Hinde, H. (1901), *The Last of the Masai* (London: Heinemann).

84. Roosevelt, T. (1910), *African Game Trails* (London: John Murray).

85. Percival, A. B. (1924), *A Game Ranger's Note Book* (London: Nisbet and Co.).

86. Patterson, J. H. (1910), *In the Grip of the Nyika* (London: MacMillan and Co.). In 1944 Osa Johnson gave a lecture at the San Diego Zoo, hosted by Ken Stott. At the end of the lecture J. H. Patterson, then an old man and retired in California, introduced himself to the two of them. He had lunch with Ken Stott the next day and confirmed the lack of buffalo in the first decade. Ken Stott reported this to me in a phone conversation in September 1981.

87. Audrey Moore visited me at Banagi with her son in 1969 to see the house she had lived in with her husband for ten years during 1930–40. She confirmed the scarcity of buffalo in the central and western Serengeti in the 1930s. See Moore, A. (1938), *Serengeti* (London: Country Life).

88. Talbot, L. M., and Talbot, M. H. (1963), 'The Wildebeest in Western Masailand', *Wildlife Monographs*, no. 12 (Washington, DC: The Wildlife Society).

89. Sinclair, *The African Buffalo*; Plowright, W. (1982), 'The Effects of Rinderpest and

Rinderpest Control on Wildlife in Africa', *Symposium of the Zoological Society, London* 50: 1–28.

90. Ruminants are part of the group of cloven-hoofed ungulates. They have special divided stomachs, the first compartment being the rumen, hence their name. Plant food is stored in the rumen where it is fermented by bacteria. At intervals the animal regurgitates a bolus of plant food and chews it into smaller fragments to help the bacteria. Once the food has been fermented, the remains of it plus the bacteria are passed on down the gut for normal digestion. Zebra, horses, rhino, and others are another group of ungulates that do not ferment their food in a rumen. Instead, these non-ruminants ferment it at the end of the gut, in the colon. This may be less efficient in digestion but it allows animals to eat coarser, tougher food and pass it through quickly, something ruminants cannot do. The two groups of ungulates separated very early in their evolution.

Chapter 5: The German Era

91. Coulson, *Tanzania, a Political Economy*, p. 30.
92. White, *The Rediscovered Country*, p. 103.
93. The porter route from Mwanza came along Speke Gulf to Handajega Hill, crossed the Mbalageti River passing west of Nyakaromo, and then went across the Grumeti River near Kirawira, finally crossing the Sibora Plains to Ikoma.
94. From Turner, *My Serengeti Years*, pp. 331–3.
95. The second route came from Waso near Loliondo, due west along the northern side of the Orangi River, past Togoro kopjes and along the Rokari River to the Grumeti River at Ikoma. This route is mentioned by R. J. Cuninghame, chapter 8 in White, *The Rediscovered Country*. He referred to the Londani River, a name that still appeared on maps of East Africa in the 1950s for what is now the Rokari River, probably an alternative pronunciation.
96. Smith, G. E. (1907), 'From the Victoria Nyanza to Kilimanjaro', *The Geographical Journal* 29: 249–69.
97. From the Nguruman forests on the Rift Valley in the east, the expedition travelled south along the escarpment until they reached the land of the Sonjo tribe opposite Lake Natron. They then turned west and reached Waso. From there they continued west to the hill called Longossa (now the eastern boundary of Serengeti National Park), where they turned north past Lobo Hill. Cunninghame had to make a side trip to Fort Ikoma at this point, walking south to Togoro kopjes then west to Ikoma. Finding no supplies at Ikoma he returned to Lobo Hill following the Grumeti River. The expedition walked north to the Bologonja River and then downstream some 20 miles before crossing west to Wogakuria Hill. From there they travelled north to the Mara River, reaching it at Rhino Plain, crossed it a few miles east of Kogatende, and conducted a loop around the Lamai wedge before returning to Rhino Plain. They then headed west, staying south of

the Mara River until they reached the Ikorongo Hills. White spent several weeks travelling south towards Ikoma and Isenye before turning north-west and ending up at Musoma, a new port on Lake Victoria only eight months old and with one building. White, *The Rediscovered Country*, p. 16.

98. White, *The Rediscovered Country*, p. 16.

99. White, *The Rediscovered Country*, p. 117.

100. Percival, A. B. (1928), *A Game Ranger on Safari* (London: Nisbet and Co.).

101. White, *The Rediscovered Country*, p. 323. Simpson went to the Serengeti plains sometime in 1914 and reported the abundance of ungulates and lions for the first time.

102. Charles Miller describes the remarkable campaign of the German general von Lettow-Vorbeck to escape the British and tie up large numbers of troops that lasted the whole war period. Miller, C. (1974), *Battle for the Bundu: The First World War in East Africa*. (London: Macdonald and Co.), London.

103. White, *The Rediscovered Country*, p. 109 and map.

Chapter 6: The Beginning of Serengeti

104. Details of the early professional hunters are given in Herne, *White Hunters*.

105. Eastman, *Chronicles of an African Trip*.

106. Martin Johnson and Osa Johnson set out on a series of photographic adventures to New Guinea, Borneo, and Africa over the years 1922–37. They recorded their exploits in a number of books. On 12 January 1937 Martin Johnson was killed in an airplane crash at Burbank, California, but Osa survived. She took no further expeditions but wrote her memoires. She died in 1953. Their biography can be found in Stott, Kenhelm W. (1978), *Exploring with Martin and Osa Johnson* (Chanute, KS: Martin and Osa Johnson Museum Press). Stott himself had met both the Johnsons and came to know Osa well. I talked with him in 1981 about the Serengeti expeditions.

107. Baumann, *Durch Massailand zur Nilquelle*; Johnson, M. (1928), *Safari* (New York: G. P. Putnam's and Sons).

108. The Eastman expedition sent three trucks to Narok, in Kenya, to bring down 40 Lumbwa tribesmen so that they could film the tribe on a lion hunt using their spears. They moved to Simpson's Camp (Seronera) on 7 August to do this. The Lumbwa also speared a buffalo bull. On the 16th Eastman, leaving the Johnsons behind until September, moved to the Bologonja spring, which they called Boundary Camp, 14 miles north of Klein's Camp. The presence of the Lumbwa tribe in Serengeti provided a mystery. None of the local tribes recognized the name or the shields from photographs. Their presence was not explained by Martin Johnson, who was only interested in describing and filming the spectacle. Indeed his books are generally short on detail for dates, locations, and ecology. George Eastman records that the Lumbwa had been

especially brought in from Narok, Kenya, to provide a show. This is also mentioned by Turner (*My Serengeti Years*). They were then transported back again. The Lumbwa are Nilotics related to the Maasai and Luo. The issue was whether these people lived in Serengeti. In fact no tribes lived in the areas that the Johnsons visited, including Moru, Mbalageti, Seronera, Banagi, and Kuka. Eastman, *Chronicles of an African Trip*.

109. Johnson, *Lion*.

110. Martin Johnson commented on this journey: 'the rock piles and brush heaps, the ravines and morasses that...we drive through are still a nightmare in my memory.' It is still the same today. Johnson, *Lion*, p. 190.

111. The papers, films, and photographs of the Johnsons are archived at the Martin and Osa Johnson Safari Museum, Chanute, Kansas, USA. On 21 July 1928 at Seronera they saw 'giraffe, zebra, impala, topi, kongoni, Tommies, Granties, dik-dik, waterbuck, bushbuck, reedbuck, bat-eared foxes, lions, cheetahs, hyena and many other species'. See Johnson, *Lion*, p. 240,

112. The diaries from the Lieurance expedition were transcribed and given to me by the grandchildren of Art Lieurance (one of three brothers), Catherine Sease, John Sease, and Ann Tiplady, wife of John. They also provided movie film for analysis of the habitats at that time. Catherine and John's mother, Mary, was on the same trip, aged ten years.

113. Johnson, M. (1935), *Over African Jungles* (New York: Harcourt, Brace); Stott, *Exploring with Martin and Osa Johnson*.

114. The Johnsons also mention numerous hartebeest on the Ndoho plains in 1933, but since this species does not occur that far west and the only photos are of topi, I presume this name was a mistake for topi.

115. Stott, *Exploring with Martin and Osa Johnson*. See also n. 106 above.

116. Herne, *White Hunters*, p. 110.

117. Turner, *My Serengeti Years*.

118. This pair contributed to the early development of conservation in Serengeti by donating funds. John Owen, Director of Tanzania National Parks, 1960–70, was very successful at raising funds for the development of Serengeti. He received donations from Donald Ker, Syd Downey, and Elizabeth Sanger to construct concrete river crossings and small dams on the Seronera River for tourists, each dam named after the donor. Syd Downey's dam crosses the Seronera River on the main road to Arusha. Donald Ker's dam (in 2009 it became a bridge) is on the main road to Banagi and the corridor. Elizabeth Sanger's dam is halfway between these two on the tourist circuit opposite the Seronera Lodge.

119. Herne, *White Hunters*, p. 169.

120. Turner, *My Serengeti Years*, pp. 34–7; Johnson, *Lion*, p. 190; Paul Hoefler also took a hunting expedition in 1928, based at Kamunya Hill near Seronera. See Hoefler, P. L. (1928), *Africa Speaks* (New York: Blue Ribbon Books).

121. Herne, *White Hunters*, p. 169.

122. Wheeler, Sara (2006), *Too Close to the Sun* (London: Jonathan Cape).

123. Huxley, *Africa View*.

124. See Moore, *Serengeti*. The road that formed the eastern boundary of the game sanctuary across the plains is uncertain. Its alignment from Seronera went further west and south (compared with the road of present times), passing east of Lake Magadi and crossing the upper Mbalageti River before turning southeast to Naabi Hill. The present alignment was constructed when Seronera was built in 1960. See *The Report of the Serengeti Committee of Enquiry, 1957* (Dar es Salaam: Government Printer). Monty Moore stayed on until the end of the 1930s. Subsequent wardens through the 1940s were John Blower and Ray Hewlett but details on events during that decade are absent. Management of the western Serengeti at Banagi was facilitated by the construction of the road over Ngorongoro Crater in 1933. This allowed a road across the plains to Naabi Hill, Lake Magadi and then north to Banagi, albeit one that was impassable in the rains. Described by Mary Teare (undated MSS.Af.s.401 at Rhodes House, Oxford University), who drove in a lorry in about November 1934 across the plains to Monty Moore's house at Banagi. Prior to this the only access to Banagi was via Kilimafeza mine and the road to Fort Ikoma and Musoma.

125. Huxley, *Africa View*.

126. The 1937 Serengeti Game Sanctuary banned hunting of lion, cheetah, leopard, giraffe, rhino, buffalo, roan, hyena, and wild dog. See Moore, *Serengeti*.

127. Huxley, *Africa View*.

128. The Game Ordinance was passed in 1940. This created the National Parks in which the killing, capture, or collection of native fauna and flora was prohibited, but it was vague on human settlement. The Serengeti National Park was designated and included the headwaters of the Simiyu and Duma Rivers, and the Endulen area, all of which were later excised. At the same time the portions of the 1930 larger closed reserve not in the park ceased to be reserved. No active implementation of protection and management was undertaken until 1951 when the park was finally proclaimed. *Report of the Serengeti Committee of Enquiry, 1957*.

129. See Moore, *Serengeti*.

130. Oscar Baumann provides a map at the back of his book showing the distribution of peoples. See Baumann, *Durch Massailand zur Nilquelle*; and Map 3 in Chapter 4 of this book.

131. The Committee of Enquiry as originally announced in August 1956 was chaired by my father, Sir Ronald Sinclair of the Court of Appeal of East Africa. He realized that before any decisions could be made a scientific review was needed. Professor Pearsall from the University of London conducted the study and reported in mid-1957. By this time my father was engaged in other work. The new committee under Sir Barclay Nihill, together with Chief Kasanda Mhoja, F. J. Mustill, and Sir Landsborough Thomson, President of the Zoological Society of London, visited Seronera 28–30 June 1957. They reported in August 1957.

132. Pearsall, W. H. (1957), 'Report on an Ecological Survey of the Serengeti National Park, Tanganyika', *Oryx* 4: 71–136.

133. Grzimek, Bernhard, and Grzimek, Michael (1960), *Serengeti Shall Not Die* (London: Hamish Hamilton).

134. H. St. J. Grant, District Officer at Loliondo in the 1950s, was an expert on the history of the Maasai. He records that Maasai lived in three sections, namely the Ol Donyo Gol in the Gol Mountains, the Olduvai Wells at Lake Lagarja and along Olduvai, and a scattered group that used the western plains. Numbers within the 1951 boundaries of the Ol Donyo Gol before 1940 were 488, and during 1940–50 were 55. Numbers of the Olduvai Wells section were 409 before 1940, 93 in 1940–50, and 79 after 1950. Numbers using the western plains seasonally were 300 pre-1940, 442 during 1940–50, and 194 after 1950. His records were obtained during November 1953–March 1954. Only the last section was affected by the realignment of boundaries in 1959. From Grant, H. St. J. (1957), *A Report on Human Habitation in the Serengeti National Park* (Dar es Salaam: Government Printer).

135. Grzimek and Grzimek, *Serengeti Shall Not Die*.

136. The Committee of Enquiry recommended the addition of the Northern Extension for the reason that 'This is fly-bush country, rich in wild life and including forest animals not normally found elsewhere in the western Serengeti, and has no inhabitants.' *Report of the Serengeti Committee of Enquiry,* 1957, p. 26.

137. The fence across the Angata Kiti valley in 1964 is reported in Turner, *My Serengeti Years*, p. 102. Hans Kruuk (pers. comm.) reports that it was constructed by John Newbould.

138. Before the park was formed, Keith Thomas was the Game Department ranger who lived at Banagi (about 1940–50) after Monty Moore (1930–9). Chief Park Wardens were Ray Hewlett (Ngorongoro) and Peter Bramwell (Banagi) (1951–6), Gordon Harvey (1959–63), Sandy Field (1963–70), Steve Stevenson (1970–3), David Babu (1973–83), L. M. Ole Moirana (1984–7), B. Maragesi (1988–91), Summay (1991–4), Maragesi (1995–8), Justin Hando (1998–2006), Martin Loiboki (2007–9), and M. G. G. Mtahiko (2009–).

139. Murray Watson was the first to record the detailed movements and population dynamics of the wildebeest in Serengeti, 1962–6, as a member of the Serengeti Research Project, based at Banagi and funded by the Food and Agriculture Organization of the United Nations. He published his PhD from this research (Watson, 'The Population Ecology of the Serengeti Wildebeest').

140. Turner, *My Serengeti Years*.

141. Lee and Martha Talbot made an early visit in 1956 and later returned to make the first detailed study largely in the Mara Reserve of Kenya in 1960. See Talbot, L. M. (1956), 'Report on the Serengeti National Park, Tanganyika', 9 pp. typed, IUCN, Brussels; Talbot and Talbot, 'The Wildebeest in Western Masailand'.

Chapter 7: The Migration of Birds

142. Reg Moreau was an amateur, initially an auditor sent to Lower Egypt 1920–7. There he developed an interest in African birds which continued when he was posted as librarian to the Amani Agricultural Research Station in the montane forests of the Usambara Mountains of Tanganyika, 1928–46. He retired from the civil service and was given a post at the Edward Grey Institute, Oxford, as editor of the bird journal *Ibis*, 1947–66. He wrote many papers on African ornithology. See also Moreau, R. E. (1966), *The Bird Faunas of Africa and Its Islands* (London: Academic Press); and (1972), *The Palaearctic-African Bird Migration Systems*, which was posthumously completed and published by James Monk. I corresponded with Reg until shortly before his death in 1970.

143. Arthur Cain became famous for his ecological genetics of colour and banding polymorphisms in snails. He was one of the first to demonstrate natural selection by predators acting on a colour polymorphism. This work is now regarded as a classic. He was also interested in colour polymorphisms in birds and went on many expeditions around the world. He became connected to Serengeti when he agreed to supervise Richard Bell for his PhD research. He was a lecturer at Oxford 1947–65. He became Professor of Zoology at the University of Manchester, and then Derby Professor of Zoology at the University of Liverpool until retirement in 1989.

144. Fritz Walther had a traumatic early career, conscripted into the German army in the Second World War. He was captured by the Russians and spent many years in a Soviet POW camp. He fell in love with gazelles when his father took him to Dresden Zoo before the war. Later he set out to describe their behaviour, writing many papers and articles. He visited Serengeti in 1964–6, and 1974–5. He spent some years at Texas A&M University. See Walther, F. R. (1995), *In the Country of Gazelles* (Bloomington: Indiana University Press).

145. I continued to record bird migrations while working on the buffalo and wildebeest population changes. This work was eventually put together in Sinclair, A. R. E. (1978), 'Factors Affecting the Food Supply and Breeding Season of Resident Birds and Movements of Palaearctic Migrants in a Tropical African Savannah', *Ibis* 120: 480–97.

Chapter 8: Socialism and War—Sort of!

146. The Serengeti Research Project was set up with funds from the United Nations Food and Agriculture Organization obtained by John Owen, Director of Tanzania National Parks. It ran from 1962 to 1966. The underlying rationale was to see whether wildlife could support food production in Africa. The main researchers were J. Verschuren on rodents, H. Klingel on zebras, M. Watson on

wildebeest, R. Sachs on parasites of ungulates, R. Bell on grazing ungulates, and H. Kruuk on hyenas.

147. The Serengeti Research Institute began in 1966 and the laboratory and houses were built near Seronera in 1968. Funds were from the Fritz Thyssen Stiftung, Germany, again obtained by John Owen. The first Director was Hugh Lamprey, Deputy Director Hans Kruuk. Tumaini Mcharo was Director 1972–6. Thereafter the Institute was administered by a series of National Park wardens and administrators from the Game Department. In 1980 the Serengeti Wildlife Research Institute, a national body in charge of all research in Tanzania, was proclaimed in parliament. The first director was appointed in 1985, based in Arusha. It later changed its name to Tanzania Wildlife Research Institute.

148. Much of this account comes from Coulson, *Tanzania, a Political Economy*; and supplemented by Ng'manyo, E. S. (1998), Introduction to *Tanzania: Portrait of a Nation* (London: Quiller Press).

149. TANU stands for Tanzania African National Union. CCM is Chama Cha Mapinduzi (Party of the Revolution).

150. The central tenets of J. K. Nyerere's socialism are the same as those of Friedrich Engels' *A Communist Manifesto* (1848).

151. In 1971 those who owned buildings worth the equivalent of $10,000 ($100,000 by 2010 prices) had them taken over at prices on a declining scale depending on age, reaching zero at ten years on the grounds that they would have paid for themselves in rent by that time. Coulson, *Tanzania, a Political Economy*. On villagization, see pp. 235–62.

152. Coulson, *Tanzania, a Political Economy*. On numbers in *Ujamaa* villages see note, p. 332.

153. Coulson, *Tanzania, a Political Economy*. Agriculture, pp. 4, 185–201.

154. Coulson, *Tanzania, a Political Economy*, p. 4.

155. From first-hand accounts of wardens Steve Stevenson and Myles Turner in 1972, and also reported in Hayes, Harold T. P. (1976), 'A Reporter at Large: The Last Place', *The New Yorker*, December 6, pp. 52–133.

156. More precisely niches are created from the partitioning of resources. A useful compendium is Levin, Simon (ed.) (2009), *The Princeton Guide to Ecology* (Princeton: Princeton University Press). See the chapter by Thomas W. Schoener, 'Ecological Niche', pp. 3–13. A useful textbook on Ecology is Krebs, Charles J. (2009), *Ecology: The Experimental Analysis of Distribution and Abundance* (San Francisco: Benjamin Cummings). Limiting similarity was first discussed by MacArthur, R. H., and Levins, R. (1967), 'The Limiting Similarity, Convergence and Divergence of Coexisting Species', *American Naturalist* 101: 377–85.

157. Regulation is explained in Chapter 3, n. 17.

158. The general consensus across Africa in the mid-1960s was that elephants were destroying the woodlands of national parks. For Murchison Falls National Park, Uganda: Laws, R. M., Parker, I. S. C., and Johnstone, R. C. B. (1975), *Elephants and*

Their Habitats (London: Clarendon Press). For Tsavo National Park, Kenya: Laws, R. M. (1969), 'The Tsavo Research Project', *Journal of Reproduction and Fertility* Supplement 6: 495–531. For Kruger National Park, South Africa: Pienaar, U. de V., van Wyk, P. W., and Fairall, N. (1966), 'An Aerial Census of Elephant and Buffalo in the Kruger National Park and the Implications Thereof on Intended Management Schemes', *Koedoe* 9: 40–108. Piennar, U. de V. (1983), 'Management by Intervention: The Pragmatic Option', in N. R. Owen-Smith (ed.), *Management of Large Mammals in African Conservation Areas*, pp. 23–6 (Pretoria: Haum Educational Publishers).

159. Corfield, T. F. (1973), 'Elephant Mortality in Tsavo National Park, Kenya', *East African Wildlife Journal* 11: 339–68.

160. On 'an act of war', see Coulson, *Tanzania, a Political Economy*, pp. 309–11.

161. Jack Inglis was the first to follow individual wildebeest using radio collars and find the exact track of the animals. See Inglis, J. M. (1976), 'Wet Season Movements of Individual Wildebeests of the Serengeti Migratory Herd', *East African Wildlife Journal* 14: 17–34.

162. Modern work on the Serengeti migrants is found in Thirgood, S., Mosser, A., Borner, M.,Tham, S., Hopcraft, G., Mlengeya, T., Kilewo, M., Mwangomo, E., Fryxell, J., and Sinclair, A. R. E. (2004), 'Can Parks Protect Migratory Ungulates? The Case of the Serengeti Wildebeest', *Animal Conservation* 7: 113–20.

163. Alan Root made the film *Serengeti Shall Not Die*, highlighting the early conservation problems of Serengeti in the 1950s. The film was made for the Grzimeks in 1958–9. He later made his name as a wildlife film-maker, producing many films in the period 1970–2000. He continued to have an interest in Serengeti and is one of its great long-time supporters. He now lives in Kenya.

164. Prime Minister Trudeau of Canada, while on a visit to Tanzania in 1979, was asked by President Nyerere for funds to offset the costs of the war against Uganda. He declined saying that Canada did not pay for other countries' wars.

165. Lamprey, 'Ecological Separation of the Large Mammal Species'; Ferrar, A. A., and Walker, B. H. (1974), 'An Analysis of Herbivore/Habitat Relationships in Kyle National Park, Rhodesia', *Journal of the Southern African Wildlife Management Association* 4: 137–47.

166. Field, 'A Comparative Study of the Food Habits of Some Wild Ungulates'; Jarman, 'Seasonal Distribution of the Large Mammal Populations'; Jarman, 'The Social Organization of Antelope'; du Toit, J. T. (1990), 'Feeding-Height Stratification among African Browsing Ruminants', *African Journal of Ecology* 28: 55–82.

167. Bell, R. H. V. (1970), 'The Use of the Herb Layer by Grazing Ungulates in the Serengeti', in A. Watson (ed.), *Animal Populations in Relation to Their Food Resources*, pp. 111–23 (Oxford: Blackwell Scientific); Bell, R. H. V. (1971), 'A Grazing Ecosystem in the Serengeti', *Scientific American* 224: 86–93; Gwynne, M. D., and Bell, R. H. V. (1968), 'Selection of Vegetation Components by Grazing Ungulates in the Serengeti National Park', *Nature* 220: 390–3.

168. Even very small differences in the skull are related to the herbivores' diet. See Codron, D., Brink, J. S., Rossouw, L., Clauss, M., Codron, J., Lee-Thorp, J. A., and Sponheimer, M. (2008), 'Functional Differentiation of African Grazing Ruminants: An Example of Specialized Adaptations to Very Small Changes in Diet', *Biological Journal of the Linnean Society* 94: 755–64. Mouth sizes are examined in Janis, C. M., and Ehrhardt, D. (1988), 'Correlation of Relative Muzzle Width and Relative Incisor Width with Dietary Preference in Ungulates', *Zoological Journal of the Linnean Society* 92: 267–84; Gordon, I. J., and Illius, A. W. (1988), 'Incisor Arcade and Diet Selection in Ruminants', *Functional Ecology* 2: 15–22. See also Gordon, I. J., and Illius, A. W. (1996), 'The Nutritional Ecology of African Ruminants: A Reinterpretation', *Journal of Animal Ecology* 65: 18–28; Gagnon, M., and Chew, A. E. (2000), 'Dietary Preferences in Extant African Bovidae', *Journal of Mammalogy* 81: 490–511.

169. See du Toit, J. T., and Owen-Smith, N. (1989), 'Body Size, Population Metabolism, and Habitat Specialization among Large African Herbivores', *American Naturalist* 133: 736–40.

170. Murray, M. G. (1993), 'Comparative Nutrition of Wildebeest, Hartebeest, and Topi in the Serengeti', *African Journal of Ecology* 31: 172–7. Murray, M. G., and Brown, D. (1993), 'Niche Separation of Grazing Ungulates in the Serengeti: An Experimental Test', *Journal of Animal Ecology* 62: 380–9.

171. Along with this chapter, I outline the theory of regulation also in Chapter 3. Results from the early 1970s were published in Sinclair, A. R. E. (1974), 'The Natural Regulation of Buffalo Populations in East Africa. IV. The Food Supply as a Regulating Factor and Competition', *East African Wildlife Journal* 12: 291–311 and earlier papers.

172. Sinclair, *The African Buffalo*.

173. Direct evidence that food, through intraspecific competition, regulates populations comes from both resident and migratory ungulates: for buffalo (Sinclair, *The African Buffalo*); wildebeest (Mduma, S. A. R., Sinclair, A. R. E., and Hilborn, R. (1999), 'Food Regulates the Serengeti Wildebeest: A 40-Year Record', *Journal of Animal Ecology* 68: 1101–22.), white-eared kob (Fryxell, J. M. (1987), 'Food Limitation and the Demography of a Migratory Antelope, the White-Eared Kob', *Oecologia*, 72: 83–91), greater kudu (Owen-Smith, N. (1990), 'Demography of a Large Herbivore, the Greater Kudu *Tragelaphus strepsiceros*, in Relation to Rainfall', *Journal of Animal Ecology* 59: 893–913), red deer (Albon, S. D., Coulson, T. N., Brown, D., Guinness, F. E., Pemberton, J. M., and Clutton-Brock, T. H. (2000), 'Temporal Changes in Key Factors and Key Age Groups Influencing the Population Dynamics of Female Red Deer', *Journal of Animal Ecology* 69: 1099–110), bighorn sheep (Festa-Bianchet, M., Gaillard, J. M., and Jorgenson, J. T. (1998), 'Mass and Density-Dependent Reproductive Success and Reproductive Costs in a Capital Breeder', *American Naturalist* 152: 367–79), and soay sheep (Coulson, T., Catchpole, E. A., Albon, S. D., Morgan, B. J. T.,

Pemberton, J. M., Clutton-Brock, T. H., Crawley, M. J., and Grenfell, B. T. (2001), 'Age, Sex, Density, Winter Weather, and Population Crashes in Soay Sheep', *Science* 292: 1528–31).

174. Water buffalo were brought from Asia to the Coburg Peninsula on the north coast of Australia in the early 1800s as a food supply for a British military garrison. In 1839 the garrison was abandoned and the animals released. The swamps and eucalyptus glades were ideal habitat for this species and so the population increased over the next 100 years eventually to cover most of the north coast of the Northern Territory. Their heavy grazing severely damaged the floodplains of the northern rivers such as the Adelaide, South Alligator, and East Alligator rivers. See Spillett, Peter G. (1972), *Forsaken Settlement* (Melbourne: Lansdowne Press).

175. The population changes in water buffalo of Australia are described in Sinclair, *The African Buffalo*, pp. 274–8.

Chapter 9: Border Closure

176. It was the first edited book on research on the Serengeti and covered the period 1958 to 1977. See Sinclair, A. R. E., and Norton-Griffiths, M. (eds) (1979), *Serengeti: Dynamics of an Ecosystem* (Chicago: University of Chicago Press).

177. Vesey had a remarkable career. From 1933 to 1936 he worked on insect pests on sugar cane in Brazil, British Guyana, and the British West Indies, then on pests of coconut palms in the Seychelles, Madagascar, and coastal East Africa (1936–9). From 1939 to 1941 he was entomologist at the Rubber Research Institute in Malaya.

He was there when the Second World War broke out and the Japanese invaded the peninsula from the north. He and many others were driven southwards to Singapore, which was a British fortress, designed to withstand any attack—any attack that is from the sea. No one had envisaged an invasion from the land and so the great gun emplacements were all pointing in the wrong direction. Singapore was filled with refugees and very few boats to take them out. By the time that the Japanese were at the causeway to Singapore all the seagoing boats had left. Vesey went down to the docks on the last night before Singapore surrendered. There he found a flat-bottomed riverboat that had been commandeered by the Royal Navy. It was taking on whatever Navy personnel it could find and a few civilians. Vesey immediately came aboard and the boat left shortly after in the dead of night and with no lights. They were armed only with a few machine guns and rifles.

Morning saw them pitching alarmingly in the Straits of Malacca heading for Java which was still in Dutch hands. They ran the gauntlet of Japanese planes but survived. Then they met a Japanese submarine; they could see its periscope.

Shortly after they saw the telltale wake of a torpedo as it came directly for them. It sped underneath and disappeared. Everyone on board had rushed to the side to watch in horror, and then to the other side, an unwise procedure as it nearly overturned the boat. A second torpedo followed the path of the first with the same result. It then dawned on everyone that it is nearly impossible to torpedo a flat-bottomed riverboat.

The submarine surfaced, perplexed at its lack of success, and determined to finish off the boat with gunfire. But the machine guns on the riverboat prevented the Japanese crew from getting out of the conning tower. Eventually the submarine gave up. Presumably it thought the riverboat was too much trouble. The riverboat made it to Java and eventually to Jakarta. From there Vesey made his way to South Africa.

Vesey recounted this story to me in 1968 and mentioned that it had been reported in *Blackwoods Magazine*. The account was by Lt Norman Bell, commander of the boat. (See Bell, N. (1943), 'Singapore to Sumatra', *Blackwoods Magazine* 253: 369–84). It bore an uncanny resemblance to Alistair MacLean's *South by Java Head*, which he had never heard of.

From 1942 to 1947 he worked as entomologist at the Middle East Anti-Locust Unit in Sudan, Saudi Arabia, and Oman. From 1947 to 1949 he was senior assistant game warden in Kenya. From 1949 to 1964 he worked at the Anti-Locust Research Centre in Abercorn, Northern Rhodesia (Zambia), looking at the red locust. This pest had its breeding grounds in the wet grasslands of swamps and lakes in Zambia and Tanzania. It was during this work that he observed the sequence of grazers in the swamps and developed his idea of the grazing succession. He was an all-round naturalist and wrote many papers and reports on the vegetation of the Middle East, East and Southern Africa, and Indian Ocean islands. The Vesey-Fitzgerald's burrowing skink (*Janetaescincus veseyfitzgeraldi*) from the Seychelles is named for his discovery.

He eventually ended up as Park Warden of Arusha National Park in Tanzania, which is where I came to work with him in the 1960s and 1970s. Vesey died in Nairobi in 1974. See Vesey-Fitzgerald, D. F. (1960), 'Grazing Succession among East African Game Animals', *Journal of Mammalogy* 41: 161–72.

178. Richard Bell conducted his PhD degree on the grazing succession of ungulates in western Serengeti 1964–7. He was also Deputy Director of the Serengeti Research Institute 1974–5. He became an expert on the soil nutrients and woodland dynamics in southern Africa, working in Zambia and Botswana. See Bell, 'A Grazing Ecosystem in the Serengeti'.

179. McNaughton, 'Serengeti Migratory Wildebeest'.

180. Norton-Griffiths, M. (1973), 'Counting the Serengeti Migratory Wildebeest Using Two-Stage Sampling', *East African Wildlife Journal* 11: 135–49.

181. See 'Foreign Policy', in Coulson, *Tanzania, a Political Economy*, pp. 304–11.

182. Ole Saibull was Director of Tanzania National Parks briefly in 1971–2 after John

Owen. He became a Regional Commissioner in southern Tanzania and then went into politics. He was a Minister at the time of the border closure.

183. Tourism started in the 1960s as a circuit from Nairobi to the Mara Reserve, then Serengeti and around to Arusha before returning to Nairobi. So nearly all tourists flew into Nairobi, stayed in hotels there, and outfitted there. Kenya received the vast majority of the spin-off revenues from tourism. This situation was a continual aggravation for the Tanzanians, and it was one of the motivations for closing the border.

184. Sinclair, 'The Lunar Cycle'.

Chapter 10: One Million Wildebeest

185. Jeremy Grimsdell was the Serengeti Research Institute ecologist after Mike Norton-Griffiths left in 1973. He stayed from 1974 until the border closure forced him out at the end of 1977. He moved to South Africa.

186. Mary Leakey continued her excavations at Olduvai after the death of her husband in 1972. She was a relative of Mike Norton-Griffiths and they met in Nairobi from time to time.

187. The footprints were discovered by Mary and colleagues in 1976. They included prints from many species of large mammals.

188. In 1973 wildebeest numbers were 700,000 and Thomson's gazelle 600,000. By 1977 the wildebeest had doubled in number (1.4 million) while the gazelle had halved (300,000).

189. See discussion in chapter 9 of S. J. McNaughton's study of wildebeest and Thomson's gazelle grazing patterns. See McNaughton, Chapter 2, note 7.

190. Leakey, M. D., and Hay, R. L. (1979), 'Pliocene Footprints in the Laetoli Beds at Laetoli, Northern Tanzania', *Nature* 278: 317–23.

Chapter 11: Outbreak of Trees

191. For references on the debate about the effects of elephants on habitats see n. 158 (Chapter 8).

192. During 1969–73 Mike Norton-Griffiths studied the effects of fire on tree regeneration and Harvey Croze observed how elephants damaged and killed trees in Serengeti. See Norton-Griffiths, M. (1979), 'The Influence of Grazing, Browsing, and Fire on the Vegetation Dynamics of the Serengeti', in Sinclair and Norton-Griffiths (eds), *Serengeti: Dynamics of an Ecosystem*, pp. 310–52; Croze, H. (1974), 'The Seronera Bull Problem', *East African Wildlife Journal* 12: 1–27, 29–47; Croze, H., Hilman, A. K. K., and Lang, E. M. (1981), 'Elephants and Their Habitats: How Do They Tolerate Each Other?', in C. W. Fowler and T. D. Smith (eds), *Dynamics of Large Animal Populations*, pp. 297–316 (New York: Wiley).

193. I took these photos with Holly Dublin. She conducted her PhD degree with me on elephant populations and vegetation change in the Mara Reserve in the

early 1980s (see Chapter 14). Dublin, H. T. (1991). 'Dynamics of the Serengeti-Mara Woodlands: An Historical Perspective', *Forest and Conservation History* 35: 169–78.

194. The decline in area burnt was first reported by Norton-Griffiths, 'The Influence of Grazing, Browsing, and Fire', pp. 310–52.

195. For effects of trees on soil and grass layer see Treydte, A. C., Heitkonig, I. M. A., Prins, H. H. T., and Ludwig, F. (2007), 'Trees Improve Grass Quality for Herbivores in African Savannas', *Perspectives in Plant Ecology Evolution and Systematics* 8: 197–205.

196. Hugh Lamprey, Director of the Serengeti Research Institute (SRI), had set up a few sites from where he took photographs in 1966, but both the photos and the documentation were destroyed by those who took control of the Research Institute after the border closure in 1977. See also Chapter 3, n. 26.

197. Monique Borgerhoff Mulder is an evolutionary anthropologist interested in demographic and economic aspects of human social organization. She works in Tanzania on the link between conservation and development. She is a Professor in Anthropology at the University of California, Davis. She was based initially at SRI in the 1980s with her husband, Tim Caro, who studied cheetahs there, but in 1991 they both moved to Lake Katavi in southern Tanzania for their research. See Borgerhoff Mulder, M., and Coppolillo, P. (2005), *Conservation: Linking Ecology, Economics and Culture* (Princeton, NJ: Princeton University Press).

198. Whistling thorn (*Acacia drepanolobium*) curiously lives in two very different soil types, the most common and widespread being impeded drainage fine silt soils (waterlogged in the wet season), often called 'black cotton' soils, which are extensive in Serengeti. The other is stony, thin soils on steep hillsides with very good drainage, heavily exposed to hot fires in the dry season.

199. At the top we found an old rain gauge, now defunct, which had been left there since the 1940s or 1950s. The puzzle was why anyone would want to measure rain there, since it would not represent the normal rainfall for the region—hills attract rain—and someone would have to climb the hill each month. Anyway, there are no known records, presumably all lost in time.

200. Kopjes are inselbergs, outcrops of granitic rock that are the tops of ancient hills. They feature as large rocks, some are small boulders a few feet high but others can reach 50 feet above the surrounding savanna or grassland. At Moru Kopjes, 30 miles south of Seronera, on the western edge of the plains, many are several hundred yards across, almost small hills. They often have dense shrubs and trees growing amongst the rocks and are favourite places for elephants to feed. Klipspringer antelope climb on the rocks, and bush hyrax (*Heterohyrax*) and rock hyrax (*Procavia*) live in the crevasses. Caves are commonly used by leopards and by lionesses to hide newborn cubs.

201. The oribi, a small antelope that lives singly or in small family groups, is confined to the broad-leaved woodlands of the north-west Serengeti and around Lobo in

238

the north-east. Studies by Simon Mduma indicated a population of about 7,000 animals in the early 1990s. See Mduma, S. A. R., and Sinclair, A. R. E. (1994), 'The Function of Habitat Selection by Oribi in Northern Serengeti, Tanzania', *African Journal of Ecology* 32: 16–29; Mduma, 'Distribution and Abundance of Oribi, a Small Antelope', pp. 220–30.

202. See Morell, Virginia (1997), 'Counting Creatures of the Serengeti, Great and Small', *Science* 278: 2058–60; Sinclair, A. R. E. (1995), 'Equilibria in Plant–Herbivore Interactions', in Sinclair and Arcese (eds), *Serengeti II*, pp. 91–113. Sinclair, A. R. E., Mduma, S. A. R., Hopcraft, J. G. C., Fryxell, J. M., Hilborn R., and Thirgood, S. (2007), 'Long Term Ecosystem Dynamics in the Serengeti: Lessons for Conservation', *Conservation Biology* 21: 580–90. Sinclair, A. R. E., Hopcraft, J. G. C., Olff, H., Mduma, S. A. R., Galvin, K. A., and Sharam, G. J. (2008), 'Historical and Future Changes to the Serengeti Ecosystem', in Sinclair, Packer, Mduma, and Fryxell (eds), *Serengeti III: Human Impacts on Ecosystem Dynamics*, pp. 7–46.

Chapter 12: Sudan

203. Linda Maddock analysed the three years of systematic surveys organized by Mike Norton-Griffiths and provided the first quantitative description of the migration. See Maddock, L. (1979), 'The "Migration" and Grazing Succession', in Sinclair and Norton-Griffiths (eds), *Serengeti: Dynamics of an Ecosystem*, pp. 104–29.

204. Hubert Braun worked on the grasslands of Serengeti 1966–9, funded by the Netherlands Tropical Research Institute (WOTRO). See Braun, H. M. H. (1973), 'Primary Production in the Serengeti: Purpose, Methods and Some Results of Research', *Annals of the University d'Abidjan (E)* 6: 171–88.

205. For adaptations of migration see Sinclair, A. R. E. (1983), 'The Function of Distance Movements in Vertebrates', in I. R. Swingland and P. J. Greenwood (eds), *The Ecology of Animal Movement*, pp. 240–58 (Oxford: Oxford University Press). Also Milner-Gulland, E. J., Fryxell, J. M., and Sinclair, A. R. E. (eds) (2011), *Animal Migration: A Synthesis* (Oxford: Oxford University Press).

206. Fryxell, J. M. (1987), 'Seasonal Reproduction of White-Eared Kob in Boma National Park, Sudan', *African Journal of Ecology* 25: 117–24; Fryxell, 'Food Limitation and Demography of a Migratory Antelope'; Fryxell, J. M., and Sinclair, A. R. E. (1988), 'Seasonal Migration by White-Eared Kob in Relation to Resources', *African Journal of Ecology*, 26: 17–31.

207. Fryxell, J. M., and Sinclair, A. R. E. (1988), 'Causes and Consequences of Migration by Large Herbivores', *Trends in Ecology and Evolution* 3: 237–41.

208. See Kruuk, H. (1972), *The Spotted Hyena* (Chicago: University of Chicago Press).

209. Fryxell, J. M., Greever, J., and Sinclair, A. R. E. (1988), 'Why Are Migratory Ungulates So Abundant?', *American Naturalist* 131: 781–98.

210. For a description of the events surrounding the 40-mile caribou herd see Sinclair, A. R. E., Fryxell, John, and Caughley, Graeme (2006), *Wildlife Ecology, Conservation and Management*, 2nd edn, p. 173 (Oxford: Blackwell Scientific).

211. The collapse of the saiga antelope in Asia was reported in Milner-Gulland, E. J., Kholodova, M. V., Bekenov, A., Bukreeva, O. M., Grachev, I. A., Amgalan, L., and Lushchekina, A. A. (2001), 'Dramatic Declines in Saiga Antelope Populations', *Oryx* 35: 340–5.

212. When migrations are prevented there is a rapid decline in population size. Migrations are discussed in Thirgood et al., 'Can Parks Protect Migratory Ungulates?'. A review of migrations around the world is Harris et al., 'Global Decline in Aggregated Migrations'; Bolger, D. T., Newmark, W. D., Morrison, T. A., and Doak, D. F. (2008), 'The Need for Integrative Approaches to Understand and Conserve Migratory Ungulates', *Ecology Letters* 11: 63–77; Milner-Gulland, Fryxell, and Sinclair, (eds), *Animal Migration*.

Chapter 13: Coup d'état

213. These conclusions are summarized in Sinclair, A. R. E. (1979), 'Dynamics of the Serengeti Ecosystem: Process and Pattern', in Sinclair and Norton-Griffiths (eds), *Serengeti: Dynamics of an Ecosystem*, pp. 1–30.

214. More correctly niche differences are measured by 'limiting similarity', namely the ratio of distance between average niche values for two species to the width of the niche. The ratio was thought to be about 1:0. Adaptations such as body size or mouth size should be in a ratio of 1:3 to 1:5. See Schoener, 'Ecological Niche', in Levins (ed.), *The Princeton Guide to Ecology*, pp. 3–13. For limiting similarity see MacArthur and Levins, 'The Limiting Similarity, Convergence and Divergence of Coexisting Species'.

215. For diet overlap of ungulates in East Africa see Sinclair, A. R. E. (1985), 'Does Interspecific Competition or Predation Shape the African Ungulate Community?', *Journal of Animal Ecology* 54: 899–918; Hansen, R. M., Mugambi, M. M., and Bauni, S. M. (1985), 'Diets and Trophic Ranking of Ungulates of the Northern Serengeti', *Journal of Wildlife Management* 49: 823–9.

216. See Strong, D. R., Simberloff, D., Abele, L. G., and Thistle, A. B. (1984), *Ecological Communities: Conceptual Issues and the Evidence* (Princeton, NJ: Princeton University Press).

217. See Holt, Robert D. (1977), 'Predation, Apparent Competition, and the Structure of Prey Communities', *Theoretical Population Biology* 12: 197–229.

218. Direct evidence for regulation through food is given in Sinclair, A. R. E., Dublin, H., and Borner, M. (1985), 'Population Regulation of Serengeti Wildebeest: A Test of the Food Hypothesis', *Oecologia* 65: 266–8.

219. The evidence for the effect of predation shaping the behaviour of ungulates is

published in Sinclair, 'Does Interspecific Competition or Predation Shape the African Ungulate Community?'.

Chapter 14: Ivory Poaching

220. Norton-Griffiths, 'The Influence of Grazing, Browsing, and Fire'.
221. The Johnson expeditions are described in Chapter 6. Their photographs and films are archived at the Martin and Osa Johnson Museum, Chanute, Kansas. The museum kindly provided copies of those that I thought I could relocate.
222. Evidence for the long-term declines in tree populations is given in Sinclair, 'Equilibria in Plant–Herbivore Interactions', pp. 91–113; Sinclair et al., 'Historical and Future Changes to the Serengeti Ecosystem', pp. 7–46.
223. Ford, *The Role of Trypanosomiases in African Ecology*.
224. Huxley, *Africa View*. Julian Huxley toured the Shinyanga and Mwanza areas in 1929 to make a report to the British Government. He comments on tsetse fly control, including burning, on pp. 75–85.
225. Syd Downey saw the Mara Triangle as his personal hunting area. See Herne, *White Hunters*, pp. 154–7.
226. The photos by Syd Downey in 1944 and ours in 1984 can be seen in Dublin, 'Dynamics of the Serengeti-Mara Woodlands'.
227. Holly Dublin conducted her PhD research with me in the Mara Reserve 1981–5. She went on to work for the World Wildlife Fund and Chaired the African Elephant Specialist Group for the IUCN Species Survival Committee for many years.
228. See Dublin, H. T., Sinclair, A. R. E., and McGlade, J. (1990), 'Elephants and Fire as Causes of Multiple Stable States for Serengeti-Mara Woodlands', *Journal of Animal Ecology* 59: 1157-64; Dublin, H. T. (1995), 'Vegetation Dynamics in the Serengeti-Mara Ecosystem: The Role of Elephants, Fire and Other Factors', in Sinclair and Arcese (eds), *Serengeti II*, pp. 71–90.
229. Hugh Lamprey was Director of the Serengeti Research Institute, 1966–72. He was an excellent bush pilot and knew the Serengeti well. See also Chapter 3, n. 26.
230. Stan Boutin conducted his graduate work with me, studying snowshoe hares at Kluane Lake, Yukon, during 1977–83. He became a faculty member of the University of Alberta and made his name studying red squirrels at Kluane.
231. These stories were recounted by Alan and Joan Root at their house on Lake Naivasha in July 1982.
232. See Dublin, Sinclair, and McGlade, 'Elephants and Fire as Causes of Multiple Stable States for Serengeti-Mara Woodlands'; Dublin, 'Vegetation Dynamics in the Serengeti-Mara Ecosystem'.

Chapter 15: Bandits

233. Photographs of this process of dismemberment by poachers are in Sinclair, *The African Buffalo*.

234. Euan Anderson and Mark Jago were with the Animal Virus Research Institute, Pirbright in Surrey, England. The Arusha Veterinary Investigation Centre had worked with me for many years on diseases of buffalo in the 1960s, and on rinderpest antibodies in 1986–7. Our main colleague was Dr Titus Mlengaya, a young vet who helped immobilize buffalo in 1986. He later became the vet for Serengeti National Park until 2006. He became the Member of Parliament in 2010 for Ramadi Town on Lake Victoria where we now live.

235. Evidence showing the absence of rinderpest is given in Dublin, H. T., Sinclair, A. R. E,. Boutin, S., Anderson, E. Jago, M., and Arcese, P. (1990), 'Does Competition Regulate Ungulate Populations? Further Evidence from Serengeti, Tanzania', *Oecologia* 82: 238–88.

236. The first clues for the role of predators in shaping prey behaviour are given in Sinclair, 'Does Interspecific Competition or Predation Shape the African Ungulate Community?'.

237. Peter Arcese and Gwen Jongejan worked with me on the behaviour of oribi and predation of ungulates in general during 1988–90, and then intermittently until 1996. Peter became a faculty member of the University of Wisconsin and later of the University of British Columbia. He became the Co-Director of the Centre for Applied Conservation Research in the Faculty of Forestry. See Sinclair, A. R. E., and Arcese, P. (1995), 'Population Consequences of Predation Sensitive Foraging: The Serengeti Wildebeest', *Ecology* 76: 882–91; Jongejan, G., Arcese, P., and Sinclair, A. R. E. (1991), 'Growth, Size, and the Timing of Births in an Individually Identified Population of Oribi', *African Journal of Ecology* 29: 340–52; Arcese, Jongejan, and Sinclair, 'Behavioural Flexibility in Small African Antelope'; Arcese, 'Harem Size and Horn Symetry in Oribi'.

238. In October 1988 while out with friends from the Canadian Wheat Scheme, Bob and Kathy Gillis, we saw the first rhino in northern Serengeti for 15 years.

239. These descriptions come from Mduma, S. A. R. (December 1989), *A Report on Bandits Attack in Kogatende Research Camp, Serengeti National Park*. Personal comminucation to the Department of Zoology, University of Dar es Salaam, Tanzania.

240. A rondavel is a small round house, usually a single room with perhaps a bathroom attached. At Kogatende we had two, one as a bedroom, the other for cooking, eating, and research work. There were also two other houses for students and staff.

241. Sinclair, A. R. E. (1995), 'Population Limitation of Resident Herbivores', in Sinclair and Arcese (eds), *Serengeti II*, pp. 194–219.

242. For what limits zebra numbers see Grange, S., Duncan, P., Gaillard, J-P., Sinclair, A. R. E., Gogan, P. J. P., Packer, C., Hofer, H., and East, M. (2004), 'What Limits the Serengeti Zebra Population?' *Oecologia* 140: 523–32.

Chapter 16: Of Princes and Polo

243. Jorie Butler Kent first formed the Friends of Maasai Mara in the mid-1980s but as she widened her mandate to include both Mara and Serengeti she changed the name to Friends of Conservation. For several years she organized her fund-raising events in February at Vero Beach, Florida.
244. Wren, P. C. (1924), *Beau Geste*. A story set before the First World War of life in the French Foreign Legion. Michael 'Beau' Geste is sent into the desert to recover a family jewel during which he performs acts of great heroism.

Chapter 17: Hando Fights Back

245. Research covering this period up to 1994 is reported in Sinclair and Arcese (eds), *Serengeti II*.
246. Lion research began with George Schaller (1966–9), then Brian Bertram (1969–73), and Jeanette Hanby and David Bygott (1974–7). Craig Packer began in 1978 and continues to this day with many students. See Schaller, G. B. (1972), *The Serengeti Lion* (Chicago: University of Chicago Press); Hanby, Bygott, and Packer, 'Ecology, Demography, and Behavior of Lions in Two Contrasting Habitats', pp. 315–31.
247. Cheetah research was begun by Brian Bertram followed by George Frame in 1974–7. In 1980 Tim Caro arrived with Monique Borghoff Mulder and conducted his fieldwork until 1983. Tony Collins and Caro's students Claire FitzGibbon and Karen Laurenson continued the long-term records for the rest of the 1980s. In 1991 Sarah Durant took over and continues to the present (2011). See Bertram, B. C. R. (1978), *Pride of Lions* (London: Dent); Frame, G. W., and Frame, L. H. (1981), *Swift and Enduring: Cheetah and Wild Dogs of the Serengeti.* (New York: E. P. Dutton); Caro, T. M. (1994), *Cheetahs of the Serengeti Plains* (Chicago: University of Chicago Press); Laurenson, M. K. (1995), 'Implications of High Offspring Mortality for Cheetah Population Dynamics', in Sinclair and Arcese (eds), *Serengeti* II, pp. 385–99.
248. The events reported here were either witnessed by myself or recounted by Justin Hando and other park staff, Simon Mduma, and other researchers in Serengeti at the time.
249. See Baumann, *Durch Massailand zur Nilquelle*.
250. Australian Rob Heinsohn worked on the lion project in the early 1990s. He went on to be lecturer at the Australian National University, Canberra.
251. The Keystone Kops were an incompetent, bumbling mob of policemen that never got their man, or only did so by mistake, featured in silent film comedies of the early twentieth century.

252. It took some six months to track them all down. Hando's agents followed them relentlessly, some as far as Dar es Salaam. There was no escape.

253. Hando left Serengeti in 2006, transferred to Ruaha National Park, and later to headquarters in Arusha, once more in charge of security.

254. Paul Tudor Jones owns the Grumeti Reserves and the Wildlife Lodge at Sasakwa. He introduced two rhino in 2007, and funded the introduction of others to Serengeti in 2010. The Frankfurt Zoological Society carried out the operation under Markus Borner. A ceremony in May 2010 was attended by President Kikwete of Tanzania and the Director of FZS, Dr Christof Schenck.

Chapter 18: Man-Eaters

255. Patterson, *The Maneaters of Tsavo*. See also Kerbis-Peterhans, J. C., and Gnoske, T. P. (2001), 'The Science of "Man-Eating" among Lions with a Reconstruction of the Natural History of the "Man-Eaters of Tsavo"', *Journal of East African Natural History* 9: 1–40.

256. The role of rinderpest and the resulting famines, sleeping sickness, and smallpox are described in Ford, *The Role of Trypanosomiases in African Ecology*.

257. Patterson, B. D., Kasiki, S. M., Selempo, E., and Kays, R. W. (2004), 'Livestock Predation by Lions and Other Carnivores on Ranches Neighboring Tsavo National Park, Kenya', *Biological Conservation* 119: 507–16.

258. Kissui, B. M. (2008), 'Livestock Predation by Lions, Leopards, Spotted Hyenas, and Their Vulnerability to Retaliatory Killing in the Maasai Steppe, Tanzania', *Animal Conservation* 11: 422–32.

259. Ikanda, D., and Packer, C. (2008), 'Ritual vs. Retaliatory Killing of African Lions in the Ngorongoro Conservation Area, Tanzania', *Endangered Species Research* 6: 67–74.

260. See n. 246 (Chapter 17). Craig Packer started research on lions with Anne Pusey in 1978. He followed on from the work of George Schaller in the 1960s, Brian Bertram in the early 1970s, and Jeanette Hanby and David Bygott in the mid-1970s. He continued this work through the difficult period of the 1980s when nearly everyone else gave up. He continues this work, now over thirty years in length.

261. Packer, C., Ikanda, D., Kissui, B., and Kushnir, H. (2005), 'Lion Attacks on Humans in Tanzania', *Nature* 436: 927–8. Also Packer, C., Ikanda, D., Kissui, B., and Kushnir, H. (2008), 'The Ecology of Man-Eating Lions in Tanzania', *Nature and Faune* 21: 10–15; Kushnir, H. (2009), 'Lion Attacks on Humans in Southeastern Tanzania; Risk Factors and Perceptions', PhD dissertation, University of Minnesota; Kushnir, H., Leitner, H., Ikanda, D., and Packer, C. (2010), 'Human and Ecological Risk Factors for Unprovoked Lion Attacks on Humans in Southeastern Tanzania', *Human Dimensions of Wildlife* 15: 5, 315–31.

262. Packer et al., 'Lion Attacks on Humans in Tanzania'. Also Packer et al., 'The Ecology of Man-Eating Lions in Tanzania'; Kushnir, 'Lion Attacks on Humans

in Southeastern Tanzania'; Kushnir et al., 'Human and Ecological Risk Factors for Unprovoked Lion Attacks'.

263. See Kissui, 'Livestock Predation by Lions, Leopards, Spotted Hyenas'; Ikanda and Packer, 'Ritual vs. Retaliatory Killing of African Lions'.

264. Between 1990 and 2003 95 per cent of the vulture population died from diclofenac poisoning on the Indian subcontinent. They now require captive breeding. See Green, R. H., Newton, I., Shutlz, S., Cunningham, A. A., Gilbert, M., Pain, D. J., and Prakash, V. (2004), 'Diclofenac Poisoning as a Cause of Vulture Population Declines across the Indian Subcontinent', *Journal of Applied Ecology* 41: 793–800.

Chapter 19: Biodiversity

265. The United Nations Conference on Environment and Development was held at Rio de Janeiro, Brazil, in 1992. Termed the 'Earth Summit' it set out a list of principles for international development and environmental cooperation, and in particular the Biodiversity Convention signed by 152 countries.

266. In the 1990s a debate developed amongst scientists as to whether the abundance of biological species in an ecosystem contributed to long-term stability of that system. It was suggested that the many species living in an ecosystem provided some sort of backup in case some species are lost by accident or outside disturbances. Many species are very similar in their role—their use of food and space and in the predator species they support—so that if one species disappears due to a disturbance then another species can take over the role. In this way the whole community of species can continue much as before, provided, of course, that not too many species are lost. If the disturbance is huge, large numbers of species are extirpated, and there are not enough left to fill the vacant niches and the ecosystem unravels. This was the theory—small disturbances can be buffered by the large diversity of species present but large disturbances cannot because too many species are lost. It was a simple and elegant theory at face value and comforting if it could be shown to be correct. Unfortunately, difficulties became apparent as soon as scientists tried to test it. First, species are all different and so none can take over another's niche exactly. Secondly, the buffering effect really depends on which species are lost because some species are more important than others in the workings of the ecosystem. If a major player is lost it is unlikely another species can replace it. The obvious example in Serengeti is if the wildebeest migration were lost, no other species can come near to filling its role and the whole ecosystem would change. On the other hand if the impala, for example, were to disappear we would guess that not much would change. In fact the buffalo population declined by some 80 per cent in the 1980s and we could not discern any major changes in how the whole ecosystem worked. So the buffering effect depends on which species are lost and not simply the number that disappear. For an

overview of some of these issues see Walker, B. H. (1992), 'Biological Diversity and Ecological Redundancy', *Conservation Biology* 6: 18–23; Walker, B. H. (1995), 'Conserving Biological Diversity through Ecosystem Resilience', *Conservation Biology* 9: 747–52; Kinzig, A. P., Pacala, S. W., and Tilman D. (eds) (2001), *The Functional Consequences of Biodiversity* (Princeton, NJ: Princeton University Press); Power, M. E., Tilman, D., Estes, J. A., Menge, B. A., Bond, W. J., Mills, S., Daily, G., Castilla, J. C., Lubchenco, J., and Paine, R. T. (1996), 'Challenges in the Quest for Keystones', *BioScience* 46: 609–20; Tilman, D. (1999), 'The Ecological Consequences of Changes in Biodiversity: A Search for General Principles', *Ecology* 80: 1455–74.

267. Sinclair, Dublin, and Borner, 'Population Regulation of Serengeti Wildebeest'.

268. Ray Hilborn began working with me in 1976 on analysis of the wildebeest population dynamics; he first went to the Serengeti as a researcher in 1991. He was in the Serengeti in 1993 and 1994 during the great drought, and has spent one to two months in the Serengeti six times since then. He is a professor at the University of Washington in Seattle, and a world expert on population dynamics and the impact of harvesting on fish stocks. In Serengeti he has worked primarily on the dynamics of the ungulates and the intensity of poaching. See Hilborn, Ray, and Hilborn, Ulrike (2012), *Overfishing: What Everyone Needs to Know* (Oxford: Oxford University Press).

269. The Serengeti Biodiversity Program was set up in 1997 to monitor plant diversity, and populations of ungulates, rodents, birds, insects, and small nocturnal carnivores. Simon Mduma was the principle researcher in Serengeti and he had a small team of Tanzanians to help him. Myself, John Fryxell, Ray Hilborn, Roy Turkington, and others provided the guidance, and supervised both Tanzanian and foreign students through the 1990s and 2000s. Grant Hopcraft studied how lions chose their prey and what habitats allowed the most successful captures. Ally Nkwabi looked at how wildebeest grazing and fire affected the community of birds. Greg Sharam examined how the same factors of grazing, browsing, and fire were causing the disappearance of riverine forests. John Bukombe studied the ungulate prey and Andrew Kittle the predators in predator–prey interactions. John Mchetto has studied the insect food of agama lizards on kopjes.

We integrated with the other long-term programmes on lions (Craig Packer), cheetah (Sarah Durant), grasslands (Sam McNaughton, Mark Ritchie, Han Olff), and the disease programme (Sarah Cleaveland, Meggan Craft, Tiziana Lembo, Katie Hampson) based at the University of Glasgow headed by Dan Hayden, and at Princeton University with Andy Dobson. Of the plants, trees were monitored from photo points and grassland communities from line transects on the short grass plains.

270. See the website for Craig Packer's lion programme including the video on lion deaths due to canine distemper: <http://www.cbs.umn.edu/lionresearch>.

271. Linda Munson made the actual discovery of distemper and this was confirmed by Max Appel. See Roelke-Parker, M. E., Munson, L., Packer, C., Kock, R., Cleaveland, S., Carpenter, M., O'Brien, S. J., Pospischil, A., Hofmann-Lehmann, R., Lutz, H., Mwamengele, G. L. M., Mgasa, M. N., Machange, G. A., Summers, B. A., and Appel, M. J. G. (1996), 'A Canine Distemper Virus Epidemic in Serengeti Lions (*Panthera leo*)', *Nature* 379: 441–5 for the spread of distemper from domestic dogs to wild carnivores. The mortality caused by distemper is given in Munson, L., Terio, K. A., Kock, R., Mlengeya, T., Roelke, M. E., Dubovi, E., Summers, B., Sinclair, A. R. E., and Packer, C. (2008), 'Climate Extremes Promote Fatal Co-infections during Canine Distemper Epidemics in African Lions', *PLoS One* 3 (1–6): e2545. It was the combination of distemper with high levels of another parasite, *Babesia*, that killed the lions. The disease was also found in hyenas.

272. Simon Mduma's analysis of the wildebeest population used the drought of 1993 to test the theory of food regulation. The results are given in Mduma, Sinclair, and Hilborn, 'Food Regulates the Serengeti Wildebeest Population'.

273. The lion trends are reported in Packer et al., 'Ecological Change, Group Territoriality and Population Dynamics'.

274. Wild dogs were common throughout the woodlands of Serengeti from the 1910s to the 1960s and were often reported in the early accounts (see Chapters 4, 5, 6). The first detailed records came from George Schaller during 1966–9. The population was monitored intermittently thereafter until the last individual was seen on the Ndabaka plains by our team in 1992. The decline in numbers is thought to be due to two factors. One was the increase in top predators and the other the spread of distemper virus from the erupting populations of domestic dogs as human numbers increased. Later work established that wild dog packs existed in the pastoral areas east of the Serengeti National Park around Loliondo, west through the hills to Kuka on the park boundary, and south into Ngorongoro Conservation Area. These are the areas where Maasai kill lions and hyenas, suggesting that the wild dogs could survive because of the lower numbers of top predators. People also kill wild dogs but apparently the dogs can tolerate this mortality more than that from other predators. Although records are poor it is possible that the wild dogs were always present in the Loliondo area but scientists had not bothered to look that far while they occurred in the park itself. The Loliondo wild dogs are now being monitored carefully by members of the Tanzania Wildlife Research Institute based at Seronera. See Ginsberg, J. R., Mace, G. M., and Albon, S. (1995), 'Local Extinction in a Small and Declining Population: Wild Dogs in the Serengeti', *Proceedings of the Royal Society, London B* 262: 221–8; Kat, P. W., Alexander, K. A., Smith, J. S., and Munson, L. (1995), 'Rabies and African Wild Dogs in Kenya', *Proceedings of the Royal Society of London B* 262: 229–33.

275. For the effects of predation on rare species in Kruger National Park, South Africa see Harrington, R. N., Owen-Smith, N., Viljoen, P., Biggs, H., and Mason,

D. (1999), 'Establishing the Causes of the Roan Antelope Decline in the Kruger National Park, South Africa', *Biological Conservation* 90: 69–78.

276. Grant Hopcraft's analysis of lion captures is in Hopcraft, Sinclair, and Packer, 'Prey Accessibility Outweighs Prey Abundance'.

277. How the pattern of top-down and bottom-up control is related to body size is reported in Sinclair, A. R. E., Mduma, S. A. R., and Brashares, J. S. (2003), 'Patterns of Predation in a Diverse Predator–Prey System', *Nature* 425: 288–90.

278. How the forest system unravels is described in Sharam, Sinclair, and Turkington, 'Serengeti Birds Maintain Forests by Inhibiting Seed Predators'.

279. Changes in the bird community with grazing and burning can be found in Nkwabi, A. K., Sinclair, A. R. E., Metzger, K. L., and Mduma, S. A. R. (2010), 'Disturbance, Ecosystem Function and Compensation: The Influence of Wildfire and Grazing on the Avian Community in the Serengeti Ecosystem, Tanzania', *Austral Ecology* 36: 1–9.

280. Changes in bird diversity with agriculture are published in Sinclair, A. R. E., Mduma, S. A. R., and Arcese, P. (2002), 'Protected Areas as Biodiversity Benchmarks for Human Impacts: Agriculture and the Serengeti Avifauna', *Proceedings of the Royal Society of London, B* 269: 2401–05.

281. Peters, Blumenschine, Hay, et al., 'Paleoecology of the Serengeti-Mara Ecosystem', pp. 47–94.

282. The role of top-down processes limiting much of the herbivores both in Serengeti and around the world is presented in Sinclair, A. R. E., Metzger, Kristine, Brashares, J. S., Nkwabi, A., Sharam, G., and Fryxell, J. M. (2010), 'Trophic Cascades in African Savanna: Serengeti as a Case Study', in J. Terborgh and J. A. Estes (eds), *Trophic Cascades: Predators, Prey, and the Changing Dynamics of Nature*, pp. 255–74 (Washington, DC: Island Press); and in Estes., J. A., et al. (2011), 'Trophic Downgrading of Planet Earth', *Science* 333: 301–6.

283. Fryxell, J., Mosser, A., Sinclair, A. R. E., and Packer, C. (2007), 'Group Formation and Predator–Prey Dynamics in Serengeti', *Nature* 449: 1041–4.

284. Sinclair, Mduma, and Brashares, 'Patterns of Predation in a Diverse Predator–Prey System'.

285. At the time of writing the effects of the Southern Oscillation are in the process of being published.

Chapter 20: The Future of Conservation

286. The international response to the threat from the road can be seen from the over 1,000 press articles in 48 countries between May and December 2010.

287. For the facts on the north road see Dobson, A. P., Borner, M., Sinclair, A. R. E., + 24 others (2010), 'Road Will Ruin Serengeti', *Nature* 247: 272–4; Sinclair, A. R. E. (2010), 'Road Proposal Threatens Existence of Serengeti', *Oryx* 44: 478–9.

288. See Map 4 (Chapter 5) from S. E. White in 1913 (White, *The Rediscovered Country*).

289. Banff National Park in Canada first built a single tarmac road. Then in the 1970s they constructed another road to create a divided highway. The increased traffic and speed caused accidents with elk and moose crossing the road, and people were killed. In the end a fence was built along this highway and both underpasses and overpasses had to be constructed to allow the migrating elk to pass. The effects of the tarmac road constructed in 1972 through Mikumi National Park, Tanzania, show that animals disturbed by the road suffer proportionately lower fatalities than species that take no notice of the road. See Newmark, W. D., Boshe, J. I., Sariko, H. I., and Makumbule, G. K. (1996), 'Effects of a Highway on Large Mammals in Mikumi National Park, Tanzania', *African Journal of Ecology* 34: 15–31.

290. Sue van Rensburg reports that a 10-mile road constructed in 2002 through the Hluhluwe-iMfolozi Park in South Africa has resulted in numerous roadkills of wildlife, especially involving wild dog, now a threatened species, but also lion, leopard, buffalo, rhino, and many antelope species. Elephant have been hit and injured. Human fatalities have occurred. The road was constructed with road humps to control vehicle speeds but cars have been recorded at 75 mph. There are no fences along the whole length but barriers were placed to reduce the risk of animals crossing in dangerous places. This served to impound animals on the road and actually increased collisions. In general the Park Manager, Sihle Nxulalo, concludes that the road has become a serious problem.

In an extensive review Ana Benitez-Lopez and others report that impacts from roads and corridors include habitat loss, intrusion of edge effects in natural areas, isolation of populations, barrier effects, road mortality, and increased human access. See Benitez-Lopez, A., Alkemade, R., and Verweij, P. A. (2010), 'The Impacts of Roads and Other Infrastructure on Mammal and Bird Populations: A Meta-analysis', *Biological Conservation* 143: 1307–16.

Michelle Gadd, in a comprehensive survey of the effects of fences, shows that there are many negative effects on wildlife populations including an increase in conflicts with humans. See Gadd, M. E. (2011), 'Barriers, the Beef Industry and Unnatural Selection: A Review of the Impacts of Veterinary Fencing on Mammals in Southern Africa', in M. J. Somers and M. W. Hayward (eds), *Fencing for Conservation*, pp. 153–86 (New York: Springer). Also see Bartlam-Brooks, H. L. A., Bonyongo, M. C., and Harris, S. (2011), 'Will Reconnecting Ecosystems Allow Long-Distance Mammal Migrations to Resume? A Case Study of a Zebra *Equus burchelli* Migration in Botswana', *Oryx* 45: 210–16; Williamson, D., and Mbano, B. (1988), 'Wildebeest Mortality during 1983 at Lake Xau, Botswana', *African Journal of Ecology* 26: 341–4.

291. Modelling of the effect of a fence preventing migrants reaching their dry season refuge is reported in Holdo, R. M., Fryxell, J. M., Sinclair, A. R. E., Dobson, A., and Holt, R. D. (2010), 'Predicted Impact of Barriers to Migration on the Serengeti Wildebeest Migration', *PLoS One* 6(1): e16370. General declines of migratory populations of mammals and birds are discussed in Bolger et al.,

'The Need for Integrative Approaches'; Harris et al., 'Global Decline in Aggregated Migrations'.

292. Myles Turner reports the construction and collapse of the fence at the western entrance of the Angata Kiti valley in 1964. Turner, *My Serengeti Years*, p. 102.

293. The full text of the letter reads:

THE UNITED REPUBLIC OF TANZANIA

MINISTRY OF NATURAL RESOURCES AND TOURISM

Director   22nd June 2011

World Heritage Centre

7, Place de Fontenoy

75352 Paris 07 SP,

FRANCE

Re: STATE OF CONSERVATION OF SERENGETI NATIONAL PARK

The United Republic of Tanzania is honoured to take this opportunity to clarify on the proposed tarmac road in northern Tanzania.

The proposed road will be constructed in two sections:

The eastern stretch of 214 km tarmac road which will be constructed from Mto wa Mbu to Loliondo,

The western stretch tarmac road that will be constructed from Makutano-Natta-Mugumu, a distance of 117 km. The stretch of 12 km from Mugumu to Serengeti National Park western border plus a corresponding stretch of 57.6 km from Loliondo to Serengeti National Park eastern border will not be tarmac. The 53 km section traversing Serengeti National Park will remain gravel road and continue to be managed by TANAPA mainly for tourism and administrative purposes as it is currently.

In view of this intended plan of construction, the State Party confirms that the proposed road will not dissect the Serengeti National Park and therefore will not affect the migration and conservation values of the Property. The ongoing ESIA will take into consideration these developments and will be submitted to WHC accordingly.

This decision has been reached in order to address the increasing socio-economic needs of the rural communities in Northern Tanzania while safeguarding the Outstanding Universal Value (OUV) of Serengeti National Park. The Government of the United Republic of Tanzania is also seriously considering the construction of a road from Mugumu to Arusha running south of Ngorongoro Conservation Area and Serengeti National Park.

Please accept, Sir, the assurance of my highest considerations.

Signed. EZEKIEL MAIGE

MINISTER FOR NATURAL RESOURCES AND TOURISM

294. Hopcraft, J. G. C. (2011), *Connecting Northern Tanzania: A Socio-economic Comparison of the Alternative Routes for a Highway from Arusha to Musoma*, report for the Frankfurt Zoological Society, Seronera, Tanzania.

295. *Nature* editorial (2010), 'An Alternative Route: A Proposed Road through the Serengeti Can Be Halted Only by Providing a Viable Substitute, Not by Criticism', *Nature* 467: 251–2.

296. The plan for the rehabilitation of the Mau forests: Government of Kenya (2010), 'Rehabilitation of the Mau Forest Ecosystem. Prepared by the Interim Coordinating Secretariat, Office of the Prime Minister, on Behalf of the Government of Kenya, with Support from the United Nations Environment Programme'.

297. For the decline in the flow of the Mara River see Gereta, E., Wolanski, E., Borner, M., and Serneels, S. (2002), 'Use of an Ecohydrological Model to Predict the Impact on the Serengeti Ecosystem of Deforestation, Irrigation and the Proposed Amala Weir Water Diversion Project in Kenya', *Ecohydrology and Hydrobiology* 2: 127–34; Gereta, E., Mwangomo, E., Wakibara, J., Wolanski, E. (2009), 'Ecohydrology as a Tool for the Survival of the Threatened Serengeti Ecosystem', *Ecohydrology and Hydrobiology* 9: 115–24.

298. I give a more in-depth analysis of the problems of community-based conservation in Sinclair, A. R. E. (2008), 'Integrating Conservation in Human and Natural Ecosystems', in Sinclair, Packer, Mduma, and Fryxell (eds), *Serengeti III*, pp. 471–95.

299. For examples of programmes designed to exploit wildlife see Child, B. (1996), 'The Practice and Principles of Community-Based Wildlife Management in Zimbabwe: The CAMPFIRE Programme', *Biodiversity and Conservation* 5: 369–98; Child, B. (ed.) (2004), *Parks in Transition* (London: Earthscan); Hulme, D., and Murphree, M. (eds) (2001), *African Wildlife and Livelihoods* (Oxford: James Currey).

300. The saiga antelope was once extremely abundant on the steppes of Asia, possibly numbering in the millions. After the collapse of the Soviet Union in 1990 a free-for-all of hunting ensued and the population declined to a few thousand in Russia and Kazakhstan. This collapse was reported in Milner-Gulland et al. 'Dramatic Declines in Saiga Antelope Populations', 340–5.

301. For examples of unsustainable use see Holmern, T., Roskaft, E., Mbaruka, J., Mkama, S. Y., and Muya, J. (2002), 'Uneconomical Game Cropping in a Community-Based Conservation Project outside the Serengeti National Park, Tanzania', *Oryx* 36: 364–72; Lybbert, T., and Barrett, C. B. (2004), 'Does Resource Commercialization Induce Local Conservation? A Cautionary Tale from Southwestern Morocco', *Society and Natural Resources* 17: 413–30; McShane, T. O., and Wells, M. (eds) (2004), *Getting Biodiversity Projects to Work: Towards More Effective Conservation and Development* (New York: Columbia University Press).

302. See Fa, J. E., Albrechtsen, L., and Brown, D. (2007), 'Bushmeat: The Challenge of Balancing Human and Wildlife Needs in African Moist Tropical Forests', in D. MacDonald and K. Service (eds), *Key Topics in Conservation Biology*, pp. 206–21 (Oxford: Blackwell Scientific).

303. For examples of the effects of agriculture on declines in bird populations see Chamberlain, D. E., Fuller, R. J., Bunce, R. G. H., Duckworth, J. C., and Shrubb, M. (2000), 'Changes in the Abundance of Farmland Birds in Relation to the Timing of Agricultural Intensification in England and Wales', *Journal of Applied Ecology* 37: 771–88; Gregory, R. D., Noble, D. G., and Custance, J. (2004), 'The State of Play of Farmland Birds: Population Trends and Conservation Status of Lowland Farmland Birds in the United Kingdom', *Ibis* 146 (Suppl. 2): 1–13; Smith, H. G. (2010), 'Consequences of Organic Farming and Landscape Heterogeneity for Species Richness and Abundance of Farmland Birds', *Oecologia* 162: 1071–9.

304. When there is a loss of rare species and an increase of a few common species the biological community has lower biodiversity. Technically this is called 'homogenization'. See Lockwood, J. L., and McKinney, M. L. (2001), *Biotic Homogenization* (New York: Kluwer Academic/Plenum).

305. Consequences of losing species at the top levels of predators are given in Estes et al., 'Trophic Downgrading of Planet Earth'; also Terborgh, J., and Estes, J. A. (eds) (2010), *Trophic Cascades: Predators, Prey, and the Changing Dynamics of Nature* (Washington, DC: Island Press).

306. For changes in Ngorongoro Crater see Runyoro, V. A., Hofer, H., Chausi, E. B., and Moehlman, P. D. (1995), 'Long-Term Trends in the Herbivore Populations of the Ngorongoro Crater, Tanzania', in Sinclair and Arcese (eds), *Serengeti II*, pp. 147–68.

307. Schneider, S. H. (1997), *Laboratory Earth: The Planetary Gamble We Can't Afford to Lose*, p. 113 (New York: Basic Books).

308. See Sillero-Zubiri, C., Sukumar, R., and Treves, A. (2007), 'Living with Wildlife: The Roots of Conflict and the Solutions', and MacDonald, D. W., Collins, N. M., and Wrangham, R. (2007), 'Principles, Practice and Priorities: The Quest for "Alignment"', in D. MacDonald and K. Service (eds), *Key Topics in Conservation Biology*, pp. 253–70, 271–90 (Oxford: Blackwell Scientific).

309. For loss of species from protected areas, and African parks in particular, see Craigie, I. D., Baillie, J. E. M., Balmford, A., Carbone, C., Collen, B., Green, R. E., and Hutton, J. M. (2010), 'Large-Mammal Population Declines in Africa's Protected Areas', *Biological Conservation* 143: 2221–8; Newmark, W. D. (2008), 'Isolation of African Protected Areas', *Frontiers in Ecology and Environment* 6: 321–8; Newmark, W. D. (1996), 'Insularization of Tanzanian Parks and the Local Extinction of Large Mammals', *Conservation Biology* 10: 1549–56; Newmark, W. D. (1995), 'Extinction of Mammal Populations in Western North American National Parks', *Conservation Biology* 9: 512–26; Newmark, W. D. (1993), 'The Role and Design of Wildlife Corridors with Examples from Tanzania', *Ambio* 12: 500–4; Newmark, W. D. (1987), 'A Land Bridge Island Perspective on Mammalian Extinctions in Western North American Parks', *Nature* 325: 430–2.

310. For examples of attrition of protected areas see Sinclair, A. R. E., Hik, D. S., Schmitz, O. J., Scudder, G. G. E., Turpin, D. H., and Larter, N. C. (1995),

'Biodiversity and the Need for Habitat Renewal', *Ecological Applications* 5: 579–87.

311. For some social, economic, ethical, and biological reasons for conservation see Norton, B. G. (1987), *Why Preserve Natural Variety?* (Princeton, NJ: Princeton University Press); Kellert, S. R. (1996), *The Value of Life* (Washington, DC: Island Press); Schneider, *Laboratory Earth*; Stolten, S., and Dudley, N. (eds) (2010), *Arguments for Protected Areas: Multiple Benefits for Conservation and Use* (London: Earthscan).

312. For the ethics of conservation see Norton, *Why Preserve Natural Variety?*

313. See the letter from the Tanzanian Government to the Director of the World Heritage Centre, in n. 293 above; and *Nature* editorial, 'An Alternative Route'.

314. A Serengeti Monitoring Trust Fund will need some $4 million capital to maintain the basic data collection for salaries and operating costs. The scientific staff are already trained. The Frankfurt Zoological Society has agreed to administer the funds. See <http://www.fzs.org>

INDEX